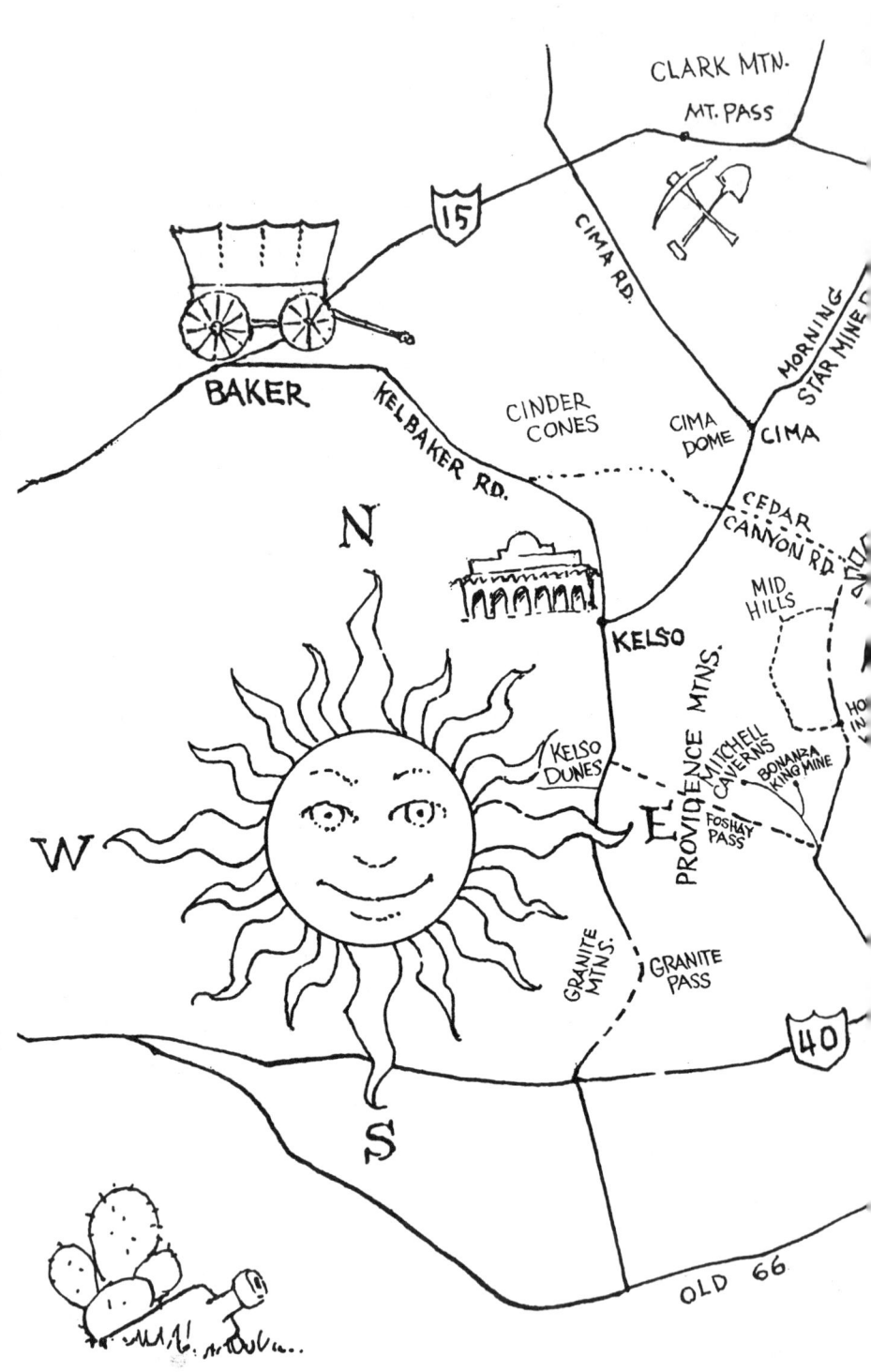

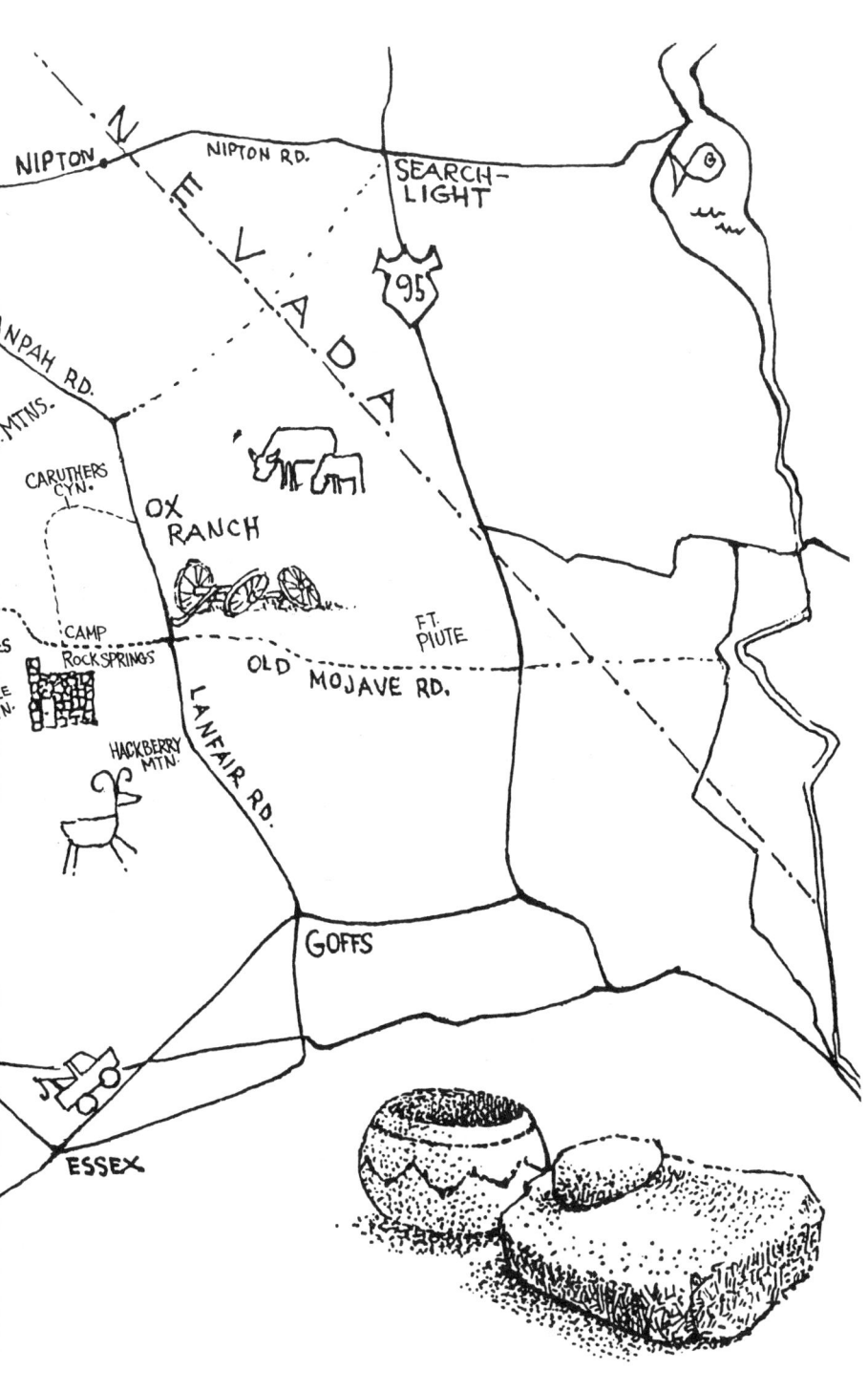

PLANTS of the EAST MOJAVE

Written by Adrienne Knute
Drawings by Carl Faber

Published by

WIDE HORIZONS PRESS
PO Box 1
Cima, CA 92323

Photographs by Adrienne Knute
Map and drawings by Carl Faber
Cover: Mojave Yucca from an acrylic painting by Carl Faber

© 1991 by Adrienne Knute. Printed in the United States of America. All rights reserved. No part of this book may be reproduced in any form or by any electronic or mechanical means, except small portions for review purposes, without the express written consent of the publisher. Published by Wide Horizons Press, PO Box 1, Cima, CA 92323.

Library of Congress Cataloging in Publication Data

Knute, Adrienne, 1936–
    Plants of the east Mojave / written by Adrienne Knute: drawings by Carl Faber.
        p.   cm.
    Includes bibliographical references and index.
    ISBN 0-938109-08-1
       1. Botany—California—East Mojave National Scenic Area. 2. Botany—California—East Mojave National Scenic Area—Pictorial works.   3. Plants—Identification.   4. East Mojave National Scenic Area (Calif.)   I. Faber, Carl, 1938–   II. Title.
QK149.K68   1991
581.9794'95—dc20   91-27773
                                                                   CIP

TABLE OF CONTENTS

Introduction to the East Mojave 1
Desert Travel 2
East Mojave Plants 3
Names of Plants 4
Organization of this Book 4
Flower Diagrams 6
Plant Discovery Walks 7
    Volcanic Rock Hill Walk 7
    Kelso Dunes Trudge 8
    Cedar Canyon Wash Walk 9
    Teutonia Peak Trail Walk 10
    Mary Beal Nature Study Trail 11
    Mojave Road-Ft. Piute Hike 12
    World's Tallest Yucca Desert Walk 13
    Caruthers Canyon Walk 13

PLANTS

Yuccas - Agaves 15
Trees 22
Cacti 35
Grasses 50
Shrubs
    White, Green flowers 54
    Yellow flowers 71
    Blue, Pink, Purple, Red flowers 104
"Wildflowers"—ephemerals and perennials
    White, Green flowers 121
    Yellow flowers 145
    Blue, Purple flowers 161
    Orange, Pink, Red flowers 176
Resources 189
Glossary 191
References—Suggested Reading 194
Plant Families 197
Plants by Flower Color 200
General Index 203

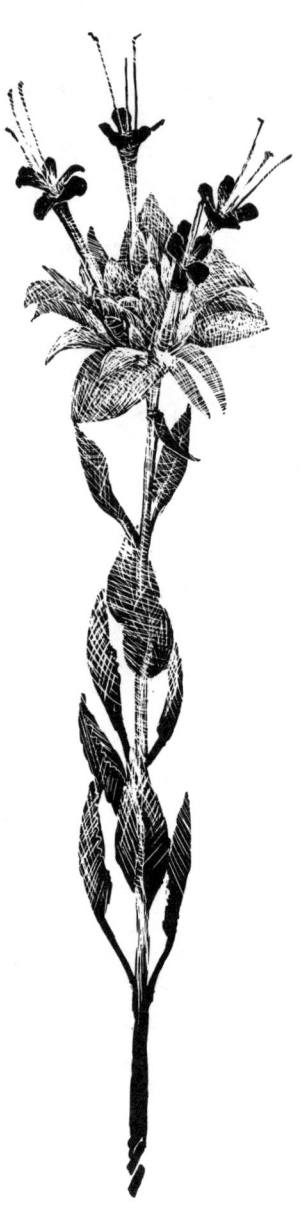

INTRODUCTION TO THE EAST MOJAVE

The East Mojave is a land of contrasts—undulating sand dunes, jagged volcanic cinder cones, scorching heat, deep snows, raging flash floods, gale-force winds, supreme solitude, full campgrounds, tiny ephemeral annual wildflowers and 300-year-old spine-tipped trees.

In 1980 1.4 million acres of the East Mojave was designated as our nation's first National Scenic Area. This land is within a few hours drive of approximately 15 million people. Here they are carried back in time to the Old West where ranching and mining are carried on as they have been for over a century.

The East Mojave National Scenic Area is supervised by the Department of Interior's Bureau of Land Management (BLM) with local offices in Needles and Barstow.

Each year more and more visitors come to enjoy the wilderness opportunities available here—camping, four-wheeling, bird watching, mountain climbing and hiking, hunting, target practice, "ghost town" exploration, rock hounding, botanizing, or just plain enjoying the scenery.

The plants of the East Mojave, along with the striking geology and native birds and animals, give this special place its unique character.

The East Mojave supports plant communities of the Great Basin Desert to the north and the Sonoran (Colorado) Desert to the south, along with its own special flora. Smoke Tree is an indicator plant of the Sonoran Desert; Joshua Tree an indicator of the Mojave Desert; Sagebrush of the Great Basin. In a 45-mile drive from Interstate 40 at Kelbaker Road to Mid-Hills Campground you will encounter all three of these indicator plants and end up in a Piñon Pine-Juniper Woodland.

I came to live at Quail Rock, at the base of Pinto Mountain near Cedar Canyon Road, in the very heart of the East Mojave National Scenic Area, in 1987. There is no other place I would rather be. Carl has been in the East Mojave since 1972, painting the desert landscapes. My main interests are the plants, animals, and birds with whom we share our lives. This love for the plants is my only qualification for writing this book. I feel so fortunate to be able to study the native vegetation, close-up, every day, I want to share the excitement, beauty and concern with others who don't have time to get to know the plants they encounter while visiting this area.

Carl and I ask you to "walk softly" while visiting our garden. It is your garden, too. Americans are so fortunate to have vast public lands, we sometimes take them too much for granted. I do not go to the city and run over rose bushes. I hope that visitors to my "front yard" (extending from Baker to Goffs) will not run over any of my 50-year-old sagebrush.

Many of the plants in the East Mojave are protected by law and should not be dug up or picked. Photographs are the best way to take plants home. There are several places where you can purchase native plants. I've had success with seeds, desert trees and shrubs sent through the mail. See back of book for listings.

Desert Travel

Be prepared! The East Mojave is not a "full-service" stop. Make sure you have everything you need before leaving the freeway. Bring lots of water.

Dirt roads may be rough or even washed-out. Some are very sandy. Telephones are scarce and sometimes not working. No police patrol the area and you might not see a BLM Ranger all day.

At Nipton you will find phones, a well-stocked little store and a beautifully-restored bed and breakfast hotel. Goffs has phones, a cafe-store and gasoline. Cima has a post office and small store. There is a phone at the Kelso Depot and one near the intersection of Cedar Canyon Road and Lanfair-Ivanpah Road. Food, gasoline and phones are available at Cima Road exit off I-15. The above services may (and do) change from time to time. Needles, Baker and Barstow are the closest towns.

There are campgrounds with minimal facilities (sometimes no water) at Mitchell Caverns, Hole-in-the-Wall and Mid-Hills. It is permissible to camp spontaneously in the desert, but hopefully you will choose a spot that has already been used for camping and leave the area clean.

The only thing predictable about the weather in the East Mojave is that it is unpredictable. Snow is common at the higher elevations in winter. Flash floods usually occur in late summer. Strong winds can kick up almost anytime. Perhaps the best time to visit, especially to see plants in bloom, is in early spring at low elevations and late spring-early summer in the mountains. Fall usually brings good weather and late blooms.

East Mojave Plants

Desert plants have devised special adaptations for life in harsh climates. The ephemerals (short-lived annuals) escape problems by growing solely under favorable conditions. Their life is short and only begins if rain has been sufficient for them to sprout, flower and produce seed. Seeds may lie dormant for many years waiting for the necessary combination of events. A "good flower year" might occur once every 5 to 15 years when all conditions have been met. Then the whole desert is almost overwhelmingly abloom.

There are many perennial plants that act much the same as annuals, except they have underground water and food storage roots or bulbs that provide new growth when conditions are right. Mariposa Lily, Giant Four O'Clock and Indian Paintbrush are perennials that send up new growth in spring, dying back in fall or winter.

Other desert plants survive all year with special adaptations such as widespread root systems, water storage tissues, waxy or hairy leaves, reduced leaves or leaf loss. Creosote Bush, all cacti and yuccas are examples of this latter type of perennial.

In the East Mojave, with its extremes in elevation, a plant that grows at 1,000 feet and also at 5,000 feet, will bloom at different times. Spring flowering in the low desert near Baker starts in February but may not occur in the higher elevations until April or May.

Soil composition is an important factor in determining what plant communities will occur. The limestone-loving plants of the Providence Mountains are not found in the decomposed granite soil of Mid-Hills.

Rainfall—and rainfall of just the right amount and at the right time of year—is the most critical influence on plant development, but other factors such as length of days, climate, temperature extremes, light intensity, etc. all play a role in the life of plants.

The growth habits listed in this book are to be used just as a general guide. So many factors influence plant height, time of bloom, leaf size, etc., it is as impossible to describe an "average" plant as it is to describe an "average" person. Even plants of the same species are variable one from another. You might find a

6-foot specimen of a plant listed as growing to a height of 3 feet, or find a February bloom on a plant normally flowering in May. But those are fun and exciting discoveries.

Names of Plants

The Latin names are listed below the common names of the plants in this book. Each Latin botanical name consists of the genus name and species name. These correspond to your last name and first name. There is often a subspecies or variety name, also. Some of these Latin names give clues to the nature of the plant such as: minutiflora-small flower, nevadensis-from Nevada; or honor a person in the botanical world: Purshia-Pursh, Gooddingii-Goodding. A plant has only one Latin name, accepted worldwide, but can have many common names. The Latin names may change, under international rules, after study proves a plant has been placed in the wrong genus. I've included many of the older, more familiar, Latin names in parentheses.

Some people don't want to be concerned with the botanical names but many are really fun to say. If you can pronounce Eucalyptus and Chrysanthemum, the Latin names for these common plants, you can surely pronounce Chrysothamnus or Isomeris. Don't be afraid of them. If you pronounce them incorrectly you'll be in the majority.

I've included some local common names that you will not find in any other plant books. These were given me by long-time residents of the East Mojave.

Organization of this Book

This book is organized by types of plants. Included are the following:

Yuccas and Agaves
Trees
Cacti
Grasses
Shrubs—Alphabetically by flower color
"Wildflowers" (annual and perennial)—Alphabetically by flower color.

A few plants do not fit neatly into the above categories so some leeway has been taken in those cases.

By no means does this book cover all of the plants in the East Mojave National Scenic Area. I've tried to include many of the common plants you will meet along with some of the unusual ones. I hope I have not left out your favorite.

Thanks

Many thanks are due John Hohstadt of Needles for his help in the field, loan of materials and manuscript suggestions; to Andrew Sanders and Oscar Clarke at the University of California, Riverside for their invaluable assistance and time; to Alan Romspert of California State University, Fullerton for careful editing of the manuscript, interesting facts, help with identification and current names; to Howard Blair for local names and plant lore; to Krista Knute, my keen-eyed daughter, for proofreading; and to Carl for patiently drawing the plants I brought home.

Flower Diagrams

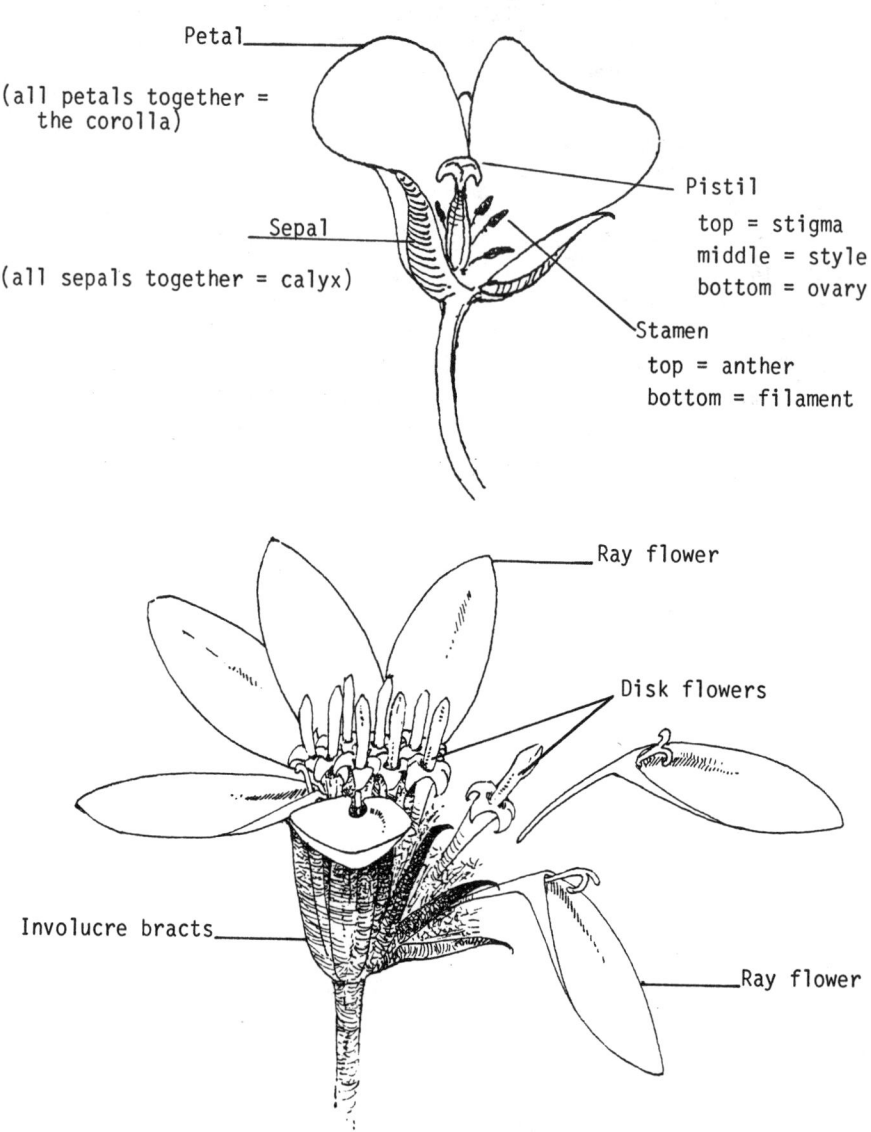

The inflorescences of the SUNFLOWER FAMILY (ASTERACEAE) are composites of many flowers. The "head" of a daisy or sunflower consists of many ray flowers on the outside and many disk flowers in the center. Each ray and each disk is an individual flower. Some members of this family have only ray flowers (dandelion) and some have only disk flowers (rabbitbrush).

PLANT DISCOVERY WALKS IN THE EAST MOJAVE

On your visit to the East Mojave there are several places which are accessible by standard cars where you can find a great variety of typical plants. Many can be seen from the window of your automobile but to really study the flora you will have to walk through the desert and examine the plants closely. A hand lens is a terrific accessory for close examinations. People often stop to photograph the showy Prickly Poppies or Sacred Daturas and discover there are hundreds of other interesting shrubs and annuals in the area which were unnoticed before walking to the large plant that caught their eye.

Volcanic Rock Hill Walk

We will start out in the low desert near the Cinder Cones—Mojave Cinder Cones National Natural Landmark—where 32 "cones" are remnants of ancient volcanic activity. Our walk takes us up a black rock volcanic hillside, through sandy washes, and across "desert pavement" of smaller volcanic rocks.

To reach this area, drive 13.2 miles east on Kelbaker Road from Baker, starting your odometer measurement from the freeway overpass. There is a parking spot for one vehicle on the west side of the hill that comes right out to the road. (Other cars would have to park farther up Kelbaker Road on the shoulder or near the wash.)

Walk around to the back side of the hill, and stay in the wash running along the back side. You will be able to see easy ways up to the top of the hill if you want to make that climb. Do not attempt it if your footing is not sure. You can see all of the plants at the base of the hill or in the surrounding desert.

You begin your visit in a little grove of Desert Willows and a few scattered Catclaws. Desert Holly, with its silvery leaves, stands out against the black rocks at the base of the hill.

While you are searching for various plants you may spy some petroglyphs near the top north side of the hill. These were pecked in the rocks by ancient Native Americans who roamed these regions. Near the petroglyphs is where Mojave Asters may be found blooming in April and May.

Follow some of the small, dry washes (watercourses) and walk on the black rock desert pavement. The following is a list of just some of the plants you will find. Many are spring annuals so, of course, will not be present at other times of the year. In April the Beavertail Cactus and Hedgehog Cactus are abloom but the Barrel Cactus and Cottontop Cactus will show their colors later in the season. If your visit is in winter, which is a good time for this low elevation, there are still plenty of plants to enjoy here.

Allscale	Desert Mallow
Anderson Lycium	Desert Trumpet
Barrel Cactus	Desert Willow
Beavertail Cactus	Fiddleneck
Bigelow Mimulus	Filaree
Bladderpod	Gold Cups
Brown-eyed Primrose	Hedgehog Cactus
Burrobush	Lilac Sunbonnet
Catclaw	Little Gold Poppy
Cheesebush	Mojave Aster
Chia	Mojave Yucca
Cottontop Cactus	Mormon Tea
Coyote Melon	Pencil Cholla
Creosote Bush	Purple Mat
Desert Chicory	Range Ratany
Desert Dandelion	Rigid Spiny-Herb
Desert Five-Spot	Skeleton Weed
Desert Holly	Wallace Daisy

Kelso Dunes Trudge

Yes, there are plants growing on the sand dunes and it's an exhilarating, heart-pounding adventure to climb the 600 feet to the top of the highest dune, then run down, heels dug in, and hear the "booms" you create on these "singing sand piles."

To reach the dunes from the town of Kelso, drive south on Kelbaker Road 7-1/2 miles to a dirt road on the west signed "Kelso Dunes—3 mi." Or, from Interstate 40 drive 14.2 miles north on Kelbaker Road to the dunes sign.

On your walk across the desert to the base of the dunes you are in Creosote Bush-Burrobush country. In spring there will be many annuals in bloom here. Gradually the sand becomes thicker and you meet those plants that have adapted to growing in shifting sands. You'll see many animal and insect tracks. There are lizards galore (Western Whiptail, Side-Blotched, Mojave Fringe-toed, etc.) and snakes. Cattle graze on the lower portions.

Plants you will meet include:

Big Galleta Grass	Burrobush
Borrego Locoweed	California Croton

Creosote Bush
Desert Lily
Desert Willow
Dune Primrose

Indian Rice Grass
Sandpaper Plant
Sand Verbena
Woolly Marigold

Dune Panic Grass—*Panicum urvilleanum*—A grass growing almost 3 feet tall with long, often exposed, creeping rhizomes, and hairy spikelets. It is almost the last plant you will meet, growing up to the slip face of the dunes.)

Cedar Canyon Wash Walk

Cedar Canyon Road crosses a large wash (a dry watercourse) 3.9 miles east of Kelso-Cima Road. There is room to park here, just off the road. If the sand looks too soft, park in the cleared areas before dropping down into the wash.

Start your walk heading north up the wash. If you want to venture up on the hillsides to the west pick spots that don't have too many chollas growing. Some people have a fear of snakes but if you stay right in the broad open wash it isn't likely you will be surprised.

This is a transition zone between Joshua Tree Woodlands, which you have just passed through, and Sagebrush Scrub, which you'll meet a few miles farther to the east.

The wash itself has a plant community of its own consisting of Desert Willow, Catclaw and Black-banded Rabbitbrush.

High on the hills across the road you'll see Utah Junipers and a few Piñon Pines.

Continue up the main wash for about 1/4 mile until you come to a side wash on the west, the entrance of which is filled with Desert Willows. The first big plant in the center of this wash is a huge-trunked Black-banded Rabbitbrush. In this side wash you will find many of the plants you have just passed, along with some new ones such as Mojave Sage, Squawbush and Woolly-Fruited Bursage.

I found the following plants growing on the slopes and the banks of these washes, on an April walk, all within about 1/4 to 1/2 mile from the parking area:

Anderson Lycium
Barrel Cactus
Beavertail Cactus
Black-Banded Rabbitbrush
Blackbrush
Blue Sage
Blue (Banana) Yucca

Buckhorn Cholla
California Buckwheat
Catclaw
Cheesebush
Chia
Cooper Lycium (Peach Thorn)
Desert Almond

Desert Mallow	Mojave Yucca
Desert Marigold	Mormon Tea
Desert Trumpet	New Mexico Thistle
Desert Willow	Palmer (Scented) Penstemon
Four Wing Saltbush	Pancake Cactus
Giant Four O'Clock	Paperbag Bush
Groundsel	Range Ratany
Hedgehog Cactus	Rattlesnake Weed
Indian Paintbrush	Rock Pea
Joshua Tree	Sacred Datura
Linear-Leaved Goldenbush	Spiny Menodora
Locoweed	Squawbush
Matchweed	Turpentine Broom
Mojave Sage	Woolly-Fruited Bursage

Teutonia Peak Trail Walk

Cima Dome is a 75-square mile eroded uplift of granite. The slope is so gradual, to really see the "dome" you must view it from some distance away. Good views can be obtained near Mid-Hills Campground. Teutonia Peak, on the east side of Cima Dome, has not eroded in the same manner as the rest of the surrounding land. It is a rugged, rocky peaklet, towering over Cima Road. The BLM, with assistance from volunteer groups, has constructed a trail leading to Teutonia Peak. To reach the trailhead from the intersection of Cima Road and Morningstar Mine Road, drive 6.3 miles north on Cima Road. There is a parking spot on the west side of the road with a sign stating "Teutonia Peak Trail - 2 mi to summit."

You will start off on a sandy road, pass through a gate, climb a stile (steps over a fence) and reach the base of a hill where the trail actually becomes a trail. All of the plants listed can be seen without climbing the peak. It is worthwhile to hike to the top of the ridge to see the rocky glens. This would be an easy backpack and a great place to spend the night.

One of the major shrubs along the trail is Spiny Menodora. It was in full bloom in early May when I walked this trail. Most Joshua Trees had finished their bloom but the Blue Yuccas were just sending up flowering stalks. Indian Paintbrush and Mojave Mound Cactus decorated the landscape with brilliant reds. Spiny Hopsage was covered with purple and green seed capsules. Many annual "belly flowers" were showing their colors. This is at a fairly high elevation so there will be flowers well into June.

Checklist of plants on Teutonia Peak Trail:

Anderson Lycium	Linear-Leaved Goldenbush
Beavertail Cactus	Locoweed
Bitterbush	Mariposa Lily
Blackbrush	Matchweed
Blue Grama Grass	Mojave Mound Cactus
Blue Sage	Mormon Tea
Blue Yucca	Old Man Cactus
Buckhorn Cholla	Pancake Cactus
California Brickellia	Paperbag Bush
California Buckwheat	Pincushion Cactus
Chia	Piñon Wormwood
Cooper Lycium (Peach Thorn)	Rock Goldenbush
Desert Almond	Silver Cholla
Desert Mallow	Spiny Hopsage
Filaree	Spiny Menodora
Forget-Me-Not	Squawbush
Four Wing Saltbush	Tansy Mustard
Giant Four O'Clock	Tidy Tips
Golden Gilia	Turpentine Broom
Hedgehog Cactus	Utah Juniper
Indian Paintbrush	Wishbone Bush
Indian Rice Grass	Wallace Daisy
Joshua Tree	Yellow Throats

Mitchell Caverns—Mary Beal Nature Study Trail

Your visit to the East Mojave will probably include a stop at Mitchell Caverns for a tour of the caves. Tours are given daily at this California State Park except during summer when the Caverns are closed. There is a small campground for overnight stays. A cactus garden has good specimens of most of the local species. At the Visitor Center you can purchase a little booklet for a quarter that identifies some of the plants in the area. The trail is one-half mile long, winding along a hillside. The booklet gives thumbnail sketches of nineteen plants and several animals. Some of the plants you might see that are not listed include:

Blackbrush	Live Forever
Canterbury Bell	Mariposa Lily
Desert Almond	New Mexico Thistle
Desert Hyacinth	Notch-Leafed Phacelia
Desert Mallow	Palmer (Scented) Penstemon
Forget-Me-Not	Rock Pea

Mojave Road—Ft. Piute Hike

If you are adventurous and an experienced hiker, a portion of the historic Mojave Road which is no longer passable to vehicles offers sweeping desert vistas culminating at shady resting spots beside one of the few year-round streams in the East Mojave. (Most of the Mojave Road has been opened up to four-wheel-drive vehicles by Friends of the Mojave Road, now headquartered in Goffs.) Along with botanical discoveries you can see the remains of an army outpost built in the 1860s, wagon ruts worn into rock when the Mojave Road was the major artery through this region, and ancient petroglyphs pecked into rocks by Native Americans.

To reach the trailhead from the intersection of Cedar Canyon Road and Lanfair-Ivanpah Road, jog north for one block to the dirt road marked with a small arrow "P T & T." Note your odometer (there may be slight variations). Head east. At a Y intersection in 3.6 miles keep right (southeast). At 9.6 miles, before the cattleguard, turn left (north). At 10.1 turn right (south). Park anywhere here or drive to the gate at 10.3. Four-wheel-drive vehicles may continue to the top of the hill where the trail starts.

An alternative starting point is 1 mile farther north, past the corral. There is a large camping area overlooking a wide gorge. You can see a trail on a hill across the canyon winding down to the gorge bottom. This, also, is a beautiful hike but the going gets tough after passing Piute Spring from the impenetrable underbrush. If you make this hike you can return via the Mojave Road and walk cross-country to your car.

The descriptions here are along the Mojave Road. I parked my car by the gate near a field of Mariposa Lilies and Filaree. The trail is very rocky so you must watch your footing at all times. It drops down to the wash at one point and eventually crosses the stream. Right before the crossing, in a red rock section, is where you will find the wagon ruts. There's usually a large rock pile beside them. You'll see petroglyphs in this area also. Be on the lookout for small rock piles pointing out the trail. Allow 1 to 2 hours each way plus time for photography, note taking and rest stops. The trail is about 2 miles long.

The stream crossing near the wagon ruts is a good place for a rest stop or lunch. There's shade, rock cliffs and sandy beaches. Other places near the streambed are clogged with plants.

Following is a list of just some of the plants I discovered on a hike in early May.

Anderson Lycium	Fremont Cottonwood
Arrowweed	Fremont Pincushion
Barrel Cactus	Ground Cherry
Beavertail Cactus	Hedgehog Cactus
Big Galleta Grass	Joshua Tree
Buckhorn Cholla	Mariposa Lily
Burrobush	Mesquite
California Buckwheat	Mojave Aster
Catclaw	Mojave Horsebrush
Chia	Mojave Yucca
Cheesebush	Mormon Tea
Creosote Bush	New Mexico Thistle
Desert Alyssium	Notch-Leafed Phacelia
Desert Chicory	Prince's Plume
Desert Hyacinth	Sacred Datura
Desert Larkspur	Salt Cedar
Desert Mallow	Virgin River Encelia
Desert Tobacco	Water-Wally
Desert Trumpet	Windmills
Desert Willow	Wishbone Bush
Filaree	Woolly-Fruited Bursage

There is a specialized plant community along Piute Creek consisting of willows, sedges, etc. Some of these plants are rare in other parts of the Scenic Area and are not dealt with in this book.

World's Tallest Yucca Desert Walk

See Mojave Yucca for directions to find a giant yucca near Hole-in-the-Wall Campground. It's an easy hike over flat desert and in a wash.

Caruthers Canyon Walk

The New York Mountains are an "island" in the desert. Many plants grow here and near Clark Mountain that are rare in other parts of the East Mojave. There is even a coastal chaparral plant community and a white fir forest—relics from an earlier, wetter era. Large Oaks, Manzanita, Coffeeberry, Silk Tassel, Holly-leaf Buckthorn, Mojave Buckbrush, and Narrow-Leaved Yerba Santa are some of the plants you may encounter here. Space in this book precludes describing the unusual flora found in the New York and Clark Mountains but you will surely want to visit these "islands in the desert."

A pleasant walk starts in Caruthers Canyon, either rock hopping along a streambed often trickling with water, or on an abandoned road, and winds up a hillside to the now-defunct Giant Ledge Mine.

To reach Caruthers Canyon from the intersection of Cedar Canyon Road and Ivanpah-Lanfair Road, drive north about 5.3 miles to the headquarters of the OX Ranch where New York Mountain Road heads west. Drive 5 1/2 miles west on N Y Mtn. Road to an unsigned dirt road. Turn north. You should soon pass a large water tank. Keep to the right at the first road junctions. In about 2 miles you will see good parking spots in places that have been used for campsites. The road gets very rocky and rutted past here. Walk up the main dirt road (which eventually becomes a trail) for a mile and a half to the mine. Keep to the left at a Y near where you left your car.

JOSHUA TREE
Yucca brevifolia var. jaegeriana
Flower: White

Agave Family - *Agavaceae*
Formerly listed in Lily Family - *Liliaceae*

It is fitting that a book on East Mojave plants should begin with the Joshua Tree. This strange tree is the symbol of the Mojave Desert. It is an indicator plant of this desert and only grows where the average annual rainfall is about 8 to 10 inches. The largest forest of Joshua Trees in the world is near Cima in the Scenic Area.

Joshua Trees received their common name from Mormon pioneers who likened them to the Biblical leader of the Israelites raising his arms to heaven.

Brevifolia is Latin for "short leaved," meaning the leaves are much shorter than those of other yuccas—only 4 to 6 inches long.

Older trees—they may be 300 years old!—are much-branched with stiff leaves tipped with spines. Leaves stick out straight but when they die they lay flat down on the branches and trunk, giving the tree a straw-thatched appearance. Leaves finally fall off the big trunk, revealing a curious checker-plated bark. There are no rings to use for dating the trees. The inside is a soft fibrous material.

There may be many blooms on a single tree. Flower clusters grow out of the tops of branches. They are made up of many waxy, succulent flowers about 2 inches long. Fruits are green pods, drying brown, filled with rows of black seeds that look like stacks of elfin poker chips. Local wildlife eagerly awaits this harvest.

The root system of the Joshuas can be studied closely since there are many fallen giants laying on the desert floor. The heavy trees fall during gale force winds that so often plague the East Mojave.

Joshua Tree and Mojave Yucca grow in the same habitats above 2,000 feet, up to about 6,000 feet elevation.

Joshua Tree with fruit

Joshua Tree bloom

BLUE YUCCA—BANANA YUCCA—SPANISH DAGGER
Yucca baccata
Flower: White

Agave Family - *Agavaceae*
Formerly listed in Lily Family - *Liliaceae*

This is the smallest of the 3 species of yucca growing in the East Mojave. It is easily identified by the bluish cast of its leaves (blades) and its low-growing habit.

Leaves are rigid with a sharp spine tip. They grow from a basal cluster to form rounded clumps 2 to 5 feet high. The margins of the leaves turn up and have fibers that peel downward, sometimes in a curl. If the leaves were flattened out they would be 2 or 3 inches wide, tapering to the upper point.

Flowers grow on a leafless stem, 2 to 3 feet long. Blooms are succulent, waxy, cream-colored, with a purple-brown tinge on the outside. Petals are about 4 inches long.

Fleshy green fruits dangle like small, fat bananas, about 6 inches long.

These yuccas, along with Mojave Yuccas, were used extensively by Native Americans. They ate the flowers, fruit and seeds. The thin, tough outer fibers were woven into ropes. Leaves were pounded to release fibers which were used for sandals and baskets. Roots supplied a soap which is still used in hair-washing ceremonies.

Many kinds of wildlife eat portions of the leaves, the flowers and, if any are allowed to grow to maturity on this easy-to-reach yucca, the fruit and seeds.

Blue Yuccas are a favorite of the range cattle. They rip off a leaf, chew the fat bottom portion, then spit it out. Many plants show evidence of this bovine practice. The flowers are the most favored delicacy of the cattle and don't last long in open range land.

East Mojave cattlemen call this yucca Oose Dagger (rhymes with Bruce). When the cattle chew the blossoms, juice oozes from their mouths. Oose is a corruption of "ooze."

Blue Yuccas often form circles made up of 3 to 5 or more plants. The inside of these circles are excellent sites for wood rat (pack rat) nests. The rats use whatever building material is handy—pine cones, little stones, cholla fruit, juniper twigs and bark, cow patties, etc. and also chew little pieces off the yucca blades—to construct their ever-expanding homes.

You'll see Blue Yuccas growing in association with Big Sagebrush and Utah Junipers in the high country of the East Mojave. They grow in Mid-Hills Campground.

Blue Yucca in bloom

MOJAVE YUCCA—SPANISH DAGGER
Yucca schidigera
Flower: White

Agave Family - *Agavaceae*
Formerly listed in Lily Family - *Liliaceae*

The majestic Mojave Yucca develops a large trunk as it matures to a usual height of 8 to 12 feet. Yellow-green, spine-tipped leaves (blades) are up to 4 feet long with peeling fiber strings along the edges. The huge clusters of cream-colored flowers are 1 to 2 feet long, growing from the center of the plant. Fruit is an oval capsule 2 to 4 inches long. Bloom is in April and May.

The yuccas have a symbiotic relationship with small pronuba moths. The moths depend on the yuccas for reproduction—they lay their eggs in the ovary of the flower and pack pollen, gathered from other yuccas, into the stigma. Thus the yucca is ensured of pollination. The larvae of the moths feed on the seeds and burrow out of the capsule when they mature. There are enough seeds left over for other animals and for the development of new seedling yuccas.

Mojave Yucca is an indicator plant of the Mojave Desert and is common in the Scenic Area, usually below 5,000 feet. Above that elevation the Blue Yucca takes over.

There is an unusual stand of Mojave Yuccas that can be reached by hiking about 1 mile off Wildhorse Canyon Road. From Hole-in-the-Wall Campground main entrance on Black Canyon Road drive south 2/10 of a mile to Wildhorse Canyon Road. Turn west for one mile to a small parking spot on the north side. A trail leads north across the desert then dips down into the big wash you've been paralleling. In about a mile you'll see "The World's Tallest Yucca!" This yucca group resembles a palm oasis. The "World's Tallest" towers above its siblings. When I first tried to estimate its height, I guessed 40 feet. Actually it is 27 feet tall, which is still mighty impressive for a Mojave Yucca.

"World's Tallest" Mojave Yucca

DESERT AGAVE—CENTURY PLANT
Agave deserti
Flower: Yellow

Agave Family - *Agavaceae*

The flower stalk of this "Century Plant" may be as high as 10 to 15 feet. The "parent" plant dies after blooming but new offsets, in a circle around the dead plant, continue growing. A Desert Agave grows from 12 to 20 years before it flowers—not a century.

Thick, succulent leaves (blades) form a large basal rosette. These leaves have spines on both margins. Flowers are up to 2 inches long, in large heads. The fruit is a dry seed pod about 2 inches long. Bloom is in late spring.

Native Americans used the leaves for fiber and baked the tender young flower stalks for food.

Desert Agave has a limited range in the East Mojave. Look for it in the Providence Mountains near Bonanza King Mine and the "ghost town" of Providence and in Foshay Pass. It also grows in the Granite Mountains.

A smaller relative, **PYGMY AGAVE** (*Agave utahensis nevadensis*) can be found in the Clark and Ivanpah Mountains. Its blades turn inward at the tips so the basal rosette of leaves is almost ball-like.

Flowering stalk of Desert Agave

ATHEL—SALT CEDAR
Tamarix aphylla
Flower: White

Tamarisk Family - *Tamaricaceae*

    These large trees are planted as windbreaks throughout the desert and are good shade trees, tolerant of poor soil. The Athel, introduced from the Mediterranean area, does not spread its seeds like its smaller relative, *Tamarix ramosissima*, so is not an invasive pest. The white flowers of summer are fairly inconspicuous.
    The railroad tracks just south of Kelso Depot are planted on both sides with Athel to protect the tracks from blow sand.

CATCLAW
Acacia greggii var. arizonica
Flower: Yellow

Pea Family - *Fabaceae*

You will immediately understand why this small tree is sometimes called "Wait a Minute Bush" when you try to untangle yourself from its small curved thorns.

Catclaw is common in and near washes in all but the higher elevations of the East Mojave. It is late to leaf out so in spring you may think it is dead. Leaves are alternate, with 4 to 7 pairs of leaflets along the sides of a central stalk. Fuzzy tiny yellow flowers hang in cylindrical spikes. Long bean pods turn a brownish red in late summer. The pods are sometimes curved and constricted between the seeds. They litter the ground beneath the tree in winter. The seeds were used by Native Americans for food.

Catclaws are often covered with thick masses of mistletoe, which may eventually kill the trees. The parasitic mistletoe that grows on Catclaw is *Phoradendron californicum*, a type that has bright red berries favored by a shiny black, crested bird—the phainopepla (silky flycatcher).

A more polite tree of the Pea Family is **MESQUITE** or **HONEY MESQUITE** (*Prosopis glandulosa*), also armed with spines but they're not curved and wicked like those on Catclaw. Mesquite is fairly rare in the Scenic Area but is, and has been, of great importance to humans and wildlife in the low deserts where it is common. Mesquite is a true water indicator plant and will send its roots down over 75 feet to tap a water supply. Along Piute Creek, where you'll find this plant, there is no need for roots so deep since water flows on the surface year round.

Mesquite grows to about 20 feet high. Small, bright green, deciduous leaves are feathery, with 7 to 17 pairs of leaflets. Tiny yellow flowers droop in spikes about 2 inches long. They are favored by bees. Long flat seed pods ripen in summer.

There is a large Mesquite tree just below Rock Springs where many Catclaws also grow.

Catclaw

Catclaw at Rock Springs

Mesquite bean

DESERT WILLOW—DESERT CATALPA
Chilopsis linearis
Flower: Pink

Bignonia Family - *Bignoniaceae*

Desert Willow is a graceful small tree, however, some plants may grow 25 feet tall. It is deciduous, so much of the year it presents an open, airy framework with the old, very long, seed pods dangling from the branches.

The leaves are often 6 inches long, very thin and tapering. Beautiful orchid-like pink flowers appear from May through summer. The narrow, 6 to 10-inch-long, seed pods burst open in autumn and spill out many flat oblong seeds, with white hairs on both ends.

Because it grows in the same habitat as our native willows and has a willow-like leaf it goes by the common name of willow. It is related to the Catalpa trees of the central and southern states.

Chilopsis means "resembling a lip" in Greek and refers to the two-lipped flowers. Linearis, as you might expect, refers to the long narrow leaves.

This plant is a cultivated landscape tree in the Southwest and, with proper irrigation, will grow 3 feet a season in its early years. It is easily grown from seed.

Desert Willow is found in the dry watercourses over much of the East Mojave below 5,000 feet. It is the only tree growing on the lower portions of the Kelso Dunes.

FREMONT COTTONWOOD
Populus fremontii
Flower: Yellow

Willow Family - *Salicaceae*

Two big Cottonwoods shade the water tank at Government Holes and can be seen from some distance away. There are many of these large trees lining Piute Creek and are a welcome sight as you hike down the old Mojave Road.

Cottonwoods grow 60 to 100 feet tall. They have deeply furrowed light bark with rough ridges. The deciduous green leaves, on 2 1/2 inch stems, are triangular and smooth and "chuckle" in the slightest breeze. They turn yellow in fall.

Male and female flowers bloom in catkins (spike-like hanging flower clusters). The female flowers develop into a string of capsules, like small beads, filled with seeds surrounded by thick white cottony hairs. When the seeds are dispersed, "cotton" covers the area.

These trees are found only at permanent water sources and transpire large amounts of water through their leaves although not as much as Salt Cedars. They provide better habitat for wildlife than Salt Cedars.

It's a treat to find a shady Cottonwood to sit under in the desert. They are widely used for landscaping throughout the West.

Cottonwood trees at Government Holes

PIÑON PINE—SINGLE-LEAVED PIÑON—NUT PINE
Pinus monophylla
Fruit: Cone

Pine Family - *Pinaceae*

Deserts and pine trees don't seem to go together, but here in the Scenic Area you'll find mountains and hillsides, particularly north-facing slopes, covered with thick forests of Piñon Pine.

This pine tree, unlike all other pines that have needles in bundles of 2 to 5, has single needles, growing out of a papery basal sheath.

Cones are round and stubby, seldom more than 2 inches long, and filled with delicious large seeds. Cones mature in August and September after the second season of growth.

The Piñon is extremely slow-growing but may reach a height of 40 feet and be 200 years old.

The nuts were a favorite of Native Americans and, because of competition with birds and animals, they picked the cones early and roasted them to release the seeds. Cones are often covered with pitch so picking them can be a sticky, messy process.

Huge flocks of piñon jays frequent the forests, their raucous cries filling the air. A jay can tell if the hard-shelled seed case contains a nut or if it's infertile and empty, so doesn't waste time cracking open those with no food value. I've found the dark chocolate-colored seeds have nuts and the light-colored ones are usually empty. Antelope ground squirrels chew up the cones to reach the sweet nuts. These seeds are nutritious and high in calories, containing about 10% protein, 54% carbohydrate, 23% fat.

Globs of pine pitch are on the branches and on the ground under the trees. It was used by Native Americans medicinally and for mending pots and baskets.

You will get an idea of how rapidly plant communities can change as you increase in elevation on Cedar Canyon Road east from Kelso-Cima Road. You begin in Creosote Bush Scrub and Joshua Tree Woodlands, travel through a brief transition zone at the wash in 4 miles, then you round a corner and meet Piñon-Juniper Woodlands on the north slopes.

You can camp under Piñon Pines at Mid-Hills Campground.

In the New York Mountains a similar nut pine, with 2 needles in a bundle, grows side by side with the Single-Leaved Piñon. This is **TWO-NEEDLE PIÑON** (*Pinus edulis*). There are good specimens in Caruthers Canyon.

Piñon Pine

Piñon Pine cone

SALT CEDAR—TAMARISK
Tamarix ramosissima
Flower: Pink or White

Tamarisk Family - *Tamaricaceae*

Salt Cedar is an introduced shrub, or small tree, which is valuable for soil erosion control but which requires so much water it has superseded many of our native plants. Seeds blown by desert winds take root easily anywhere near a water source. Environmental groups are working to check its invasion in such places as Saratoga Springs in Death Valley, 1000 Palms, and Camp Cady and Afton Canyon in the East Mojave.

It gets its common name of Salt Cedar because it can grow in highly saline soils and the tiny scalelike leaves resemble those of cedars.

In spite of its obnoxious habits, one cannot help but admire the long, graceful, flower clusters blooming in spring.

Look for this small tree on Kelso-Cima Road north of the Kelso Depot and by the Cima post office.

SMOKE TREE
Psorothamnus spinosus
Flower: Blue-purple

Pea Family - *Fabaceae*

Many desert visitors miss the mass of indigo blooms on this usually lifeless looking tree because it puts on its showy display in summer. The rest of the year the ash gray branches, ending in spines, give the appearance of clouds of smoke when viewed from a distance.

The trees can grow as high as 30 feet but without adequate moisture often die back into unsightly tangles.

Most plant books list 1,500 feet elevation as the upper limit of Smoke Trees, however, you'll find some handsome specimens at higher elevations in the East Mojave—a few near I-40 at Kelbaker Road (many more south of I-40) and lining Goffs Road heading to Needles.

These plants grow in sandy washes and depend on summer cloudbursts for moisture. Many seedlings are uprooted by the strong currents of flash floods. The hard-coated seeds will germinate only after being scarified (broken or abraded) by the tumbling action of sand and water when the washes run.

The botanical genus name for this plant has changed. It has been listed as *Parosela spinosa* and *Dalea spinosa*. The species name "*spinosus*," as you can imagine, means "spiny" or "thorny" and has remained the same except for the ending "us" which changed to reflect the genus ending change.

The few gland-dotted leaves of the tree grow in spring and are shed before flowering. The young trees have well developed leaves which they lose early in life. The purple blooms have a wide upper petal, 2 side petals, and 2 joined lower petals. The calyx is ringed with brownish-orange glands.

It's worth a visit to the lower desert areas in June or July to see this handsome tree in bloom.

*Smoke Tree at Kelbaker Road and Interstate 40.
Desert Senna in foreground*

TURBINELLA OAK—SCRUB OAK—SHRUB LIVE OAK
Quercus turbinella
Fruit: Brown acorn

Beech Family - *Fagaceae*

The canyons and slopes of Piñon-Juniper country in the East Mojave are home to this little oak. It is a large shrub, or small tree, up to 10 feet tall. Leaves are hollylike with spine-tipped margins, up to 1 inch long, shiny green on top, hairy and paler on the underside.

Male and female flowers are very small. Male flowers hang in tassels; female flowers are bud-like in a stemmed cluster. Fruit is an acorn with a "little top" (turbinella)— the acorn cup. Acorns on these plants develop in one season.

The bark of oaks and the "oak apples"—the galls commonly seen on the branches caused by a wasp larva—have high tannin content. Oak bark tea is used medicinally. Acorns were a source of food for Native Americans and are eaten by many animals.

You'll see Turbinella Oak throughout Caruthers Canyon in the New York Mountains growing along with its larger relative, **CANYON LIVE OAK** (*Quercus chrysolepis*).

New acorns forming on Turbinella Oak

UTAH JUNIPER—JUNIPER—CEDAR
Juniperus osteosperma
Fruit: Blue

Cypress Family - *Cupressaceae*

Living among the Junipers, I've developed a real love for these sturdy trees. I read under their welcome shade in summer. They brighten the landscape in winter with their deep green foliage. Carl often features their twisted, shredding trunks in his paintings. We treasure each Juniper on our property and deplore the all-too-common practice of some unthinking people of tearing off a dead limb for a campfire, leaving a raw, ugly scar. Most Junipers have dead branches which add to the character and form of the trees.

In the higher elevations of the East Mojave, above 4,000 feet, you leave the Creosote Bush Scrub plant community, pass through Joshua Tree Woodlands and enter Piñon-Juniper Woodlands and Sagebrush Scrub much like the Great Basin Desert of Nevada.

When people visit us for the first time, a common reaction is, "I thought you lived in the desert." This Juniper-Sagebrush country does not fit the picture of "desert" in the minds of people who imagine rolling sand dunes with little vegetation. We are only 25 miles from Kelso Dunes, but what a difference!

The Utah Junipers, locally called Cedars, from which Cedar Canyon takes its name, are aromatic evergreen trees, with scale-like, flattened leaves and a definite trunk, usually growing to about 15 feet, but some greatly exceeding that height. Fruit is a one-seeded bluish-gray berry.

In March great poofs of pollen float through the air when even a slight breeze rustles the trees. Tiny brown male "flowers" (cones) grow at the ends of the branches. Female cones develop into berries by a gradual union of the scales. They start out looking like very tiny, fleshy, purple pine cones.

Juniper berries are an important food source for wildlife and humans. I find the young fruit very pleasant—with a familiar gin flavoring. They can be used to season soups, stews and casseroles. Old berries may remain on the tree for over a year, drying out and becoming rock hard.

The bark of the tree shreds easily and was used by Native Americans for bedding. Wood rats (pack rats) also use the silky shredded bark as bedding in their nests.

Some trees have clusters of 1 to 10 teensy pineapple-shaped cones, at first fleshy, then drying to a hard brown. These are galls and if you cut one open you'll find a small grublike worm inside.

Early ranchers and homesteaders in this area used Juniper limbs for fence posts so many of the old trees have had their large branches cut off. Remnants of the fences are still standing throughout the Scenic Area. Lightning strikes have taken their toll, leaving dead, burned "snags." A large portion of the Juniper forest on top of Table (Top) Mountain burned in the mid-1980s leaving hundreds of dead but still-standing trees. It's worth a climb to the top to see this eerie landscape.

Many Junipers have Mistletoe (*Phoradendron juniperinum*) growing on their branches. This does not resemble our familiar Christmas decoration at all but has thin stems with minuscule leaves and straw-colored berries.

A slightly smaller but similar tree, **CALIFORNIA JUNIPER** (*Juniperus californica*) is found in the Granite Mountains.

Winter scene with Utah Juniper

BARREL CACTUS—BISNAGA
Ferocactus acanthodes var. lecontei
Flower: Yellow

Cactus Family - *Cactaceae*

The Barrel Cactus starts out life as a little round sphere and becomes cylindrical as it matures. They often lean toward the south and prefer sunny southern slopes. We call a little hill by our home "Barrel Mountain" because of the great number of these cacti on the south-facing slope. The north side of the hill has Juniper and Piñon Pine.

Barrels may be 20 or 30 years old. These striking cacti are usually 1 to 4 feet tall and are sometimes confused with young Saguaros, the cactus symbol of Arizona. The spines of the Barrel Cactus are very different, however. Saguaro's spines are straight and white, while the Barrel Cactus is armed with flat, reddish or golden spines, the central ones curving downward at the tip—quite a fierce defense system.

They do not contain "water" as the old myth suggests, but the mashed pulp might save a life if one could devise a way to cut open the heavily armed flesh. This would, naturally, kill the cactus.

This type of Barrel usually grows singly but sometimes you will find clumps of 2 to 5 or more plants.

They have thick, stout ribs but these are almost concealed by the long spines. The accordion ribbing allows the cactus stem to swell or shrink as it stores or loses water.

Yellow flowers appear in late spring and early summer. They may be 2 inches wide and grow in a ring on the top of the stem.

Barrel Cactus fruits look like miniature pineapples and are filled with tiny black seeds, greatly favored by birds and animals. If not pulled off by wildlife, the empty seed capsules will stay on the plant all year.

The common name of Bisnaga may be derived from a Nahuatl (Aztec) word roughly meaning "vicinity of thorns."

This cactus is listed in some plant books as *Echinocactus acanthodes*.

You'll see Barrel Cacti on the Cedar Canyon Wash Walk and can find them in the Cinder Cone area near Kelbaker Road east of Baker. The road leading to Mitchell Caverns is overseen by hundreds of Barrels.

Barrel Cactus on Pinto Mountain. Mojave Mound Cactus and Linear-Leaved Goldenbush in foreground. Utah Junipers in background.

BEAVERTAIL CACTUS
Opuntia basilaris var. basilaris
Flower: Magenta

Cactus Family - *Cactaceae*

Beavertail is one of the crown jewels of the desert. It is a low-growing cactus, usually not more than a foot high and is commonly overlooked until it bursts into bloom. Then, the brilliant 2 or 3-inch-wide flowers almost cover the plant in a spectacular riot of color.

This cactus has flat green to purplish pads, spineless but armed with bristles (glochids) that can require painstaking patience to remove. Don't Touch! Pads are often almost heart-shaped. In times of drought they appear quite wrinkled.

Flowers are clustered on the upper rim of the pads. The fruit is about an inch long, dry and spineless but covered with glochids.

All parts of this cactus were eaten by Native Americans—pads (scraped free of glochids), flowers, fruits and seeds.

Beavertail grows up to about 6,000 feet in the East Mojave and flowers from March to June. It will get your attention when it's in bloom; otherwise, look for it on rocky slopes and near roadsides. You'll see it on the Cedar Canyon Wash Walk and above Rock Springs.

Beavertail Cactus beside old post office/store in Kelso

BUCKHORN CHOLLA—CANE CHOLLA
Opuntia acanthocarpa var. coloradensis
Flower: Yellow

Cactus Family - *Cactaceae*

The Opuntia genus of the Cactus Family has certain recognizable characteristics even though the two types of opuntias look very different. They have jointed stems, either cylindrical like chollas (pronounced choy-ya) or flat like prickly pears, and areoles—a small defined area—with glochids (sharp, barbed bristles different from spines).

The Buckhorn Cholla of the East Mojave has many branches and, though usually 3 to 5 feet high, can grow over 8 feet. The joints are 6 to 12 inches long. Spines, with tiny glochids beneath, are straw-colored, about an inch long. Each spine is covered with a papery sheath which can easily be pulled off from the top. In spring, when new growth occurs, the tops of joints sprout soft, flexible, fleshy leaves, somewhat like pine needles. These soon drop off.

Petal color is variable but blooms are generally yellow or tinged with red or brown, up to 2 inches across. Bloom is in May and June. The fruit is dry and shriveled and very spiny. When the mature fruits drop to the ground they can be a hazard to man and beast as can any joints that may have broken off the cactus.

Bird nests are often located in the protective arms of this cholla. Wood rats use the fallen joints as a defensive "roof" for their sprawling nests.

The desert is "littered" with dead, fallen chollas. After the flesh rots away, the beautiful airy, open-holed, woody skeleton remains. The holes are the areoles and the woody part is the vascular tissue.

Buckhorn Cholla grows up to about 5,500 feet elevation and is very common throughout the East Mojave. Beautiful "forests" may be seen on Lanfair Road north of Goffs and in Woods Wash in the Woods Mountains.

Buckhorn Cholla

COTTONTOP CACTUS
Echinocactus polycephalus
Flower: Yellow

Cactus Family - *Cactaceae*

Polycephalus means "many heads." This cactus grows in clumps with 10 to 30 stems—or "heads." Each round, ribbed stem is about 12 inches high and covered with clusters of stout spines. The stems resemble young Barrel Cacti.

Cottony "wool" covers the flower base and fruit and remains on the plant all year, making it easy to identify. Flowers bloom at the top of the stems.

Cottontops grow on rocky slopes below 5,000 feet. There are good specimens in the Grotto Hills, northwest of the intersection of Cedar Canyon Road and Lanfair Road and in the Cinder Cone area off Kelbaker Road east of Baker.

HEDGEHOG CACTUS
Echinocereus engelmannii var. chrysocentrus
Flower: Magenta

Cactus Family - *Cactaceae*

The stout cylindrical stems of Hedgehog grow in groups of 2 to 10 or so, but sometimes you'll find a solitary stem. They are usually up to 1 foot tall. Many straight spines cover the ribs. The brilliant magenta blooms appear in April and May and are a photographer's delight.

The center of the flower is filled with hundreds of delicate stamens and a large pistil with thin, rich green segments at its top.

Many species of cacti have stamens that are thigmotactic (they move toward a touch). If you put your finger in the center of the flower and wiggle it, the stamens will move toward your finger. When insects land in the flower the stamens move toward them and thus rub pollen onto the insects to be transported to other flowers.

Hedgehog Cactus is fairly common in the East Mojave but, until it blooms, can go unnoticed because of its small size and habit of growing close to rocks. Look for it on the Volcanic Hill Walk and the Cedar Canyon Wash Walk. It is abundant at Mitchell Caverns.

MOJAVE MOUND CACTUS
Echinocereus triglochidiatus var. mojavensis
Flower: Scarlet

Cactus Family - *Cactaceae*

Every other day in March I walk up the hill to see if the buds on my favorite Mound Cactus have opened. It is always the first to bloom of these early flowering cacti.

The round clumps, formed by sometimes many hundreds of cylindrical stems growing tightly packed together, can be 2 feet tall and 3 yards across. Most of the mounds are smaller—about 2 feet across. Each stem produces its own flowers so a mass bloom can be exquisite.

The stems have 8 to 13 ribs; each "nipple" along the rib has an areole with a cluster of white, gray, or yellow flexible spines that interlace with those around them. New spines, on the tops of stems, appear in April and are red.

The buds burst through the flesh of the cactus on the sides of the stems. The many-petaled, bright scarlet flowers stay open for several days. They attract all kinds of insects that collect pollen from the many stamens. The stamens are reddish tipped and the stigma is divided into a number of bright green sections. If you rub the flower surface you will see it is somewhat leathery to the touch.

If you'll stop on Cedar Canyon Road anywhere near Government Holes and walk the slopes on the north side of the road, you'll find many good specimens of Mojave Mound. They like to grow around rocks. You'll also see it on the Teutonia Peak Trail.

MOJAVE PRICKLY PEAR
Opuntia phaeacantha
Flower: Yellow

Cactus Family - *Cactaceae*

Mojave Prickly Pear is a wicked-looking prostrate, sprawling plant with large, round, greenish-purple pads, 8 to 12 inches across. Areoles are spaced far apart in symmetrical rows. Spines are up to 2-1/2 inches long; the main ones stick straight out. There are many glochids.

This cactus blooms in May and June with delicate yellow flowers. Purple, spineless, edible fruit follows.

Of all the types of cacti in the Scenic Area this variety seems to be eaten by animals the most. It's often completely destroyed.

You'll find Mojave Prickly Pear in Caruthers Canyon of the New York Mountains, the Bonanza King Mine area of the Providence Mountains and—the biggest stand I've seen—near Hackberry Spring on Hackberry Mountain.

OLD MAN CACTUS—GRANDDADDY CACTUS
Opuntia erinacea var. erinacea
Flower: Yellow

Cactus Family - *Cactaceae*

I meet Old Man often, much to my chagrin, if I don't watch where I put my feet. It's the most common cactus on our property, usually not more than a foot high, and is often concealed in thick grass or bushes. Even though I'm always pulling out its spines from my shoes and toes, I still have a great affinity for it. Whenever I see a pad that has been pulled off by some animal, I replant it. The pads root easily.

The oblong or elliptical pads are green to brownish-purple. They are jointed but usually not more than 2 or 3 pads high. Areoles have 4 to 7 stiff white spines. The whole appearance is one of shaggy gray hair.

Yellow flowers, 2 inches wide, bloom in May and June. Fruit is about 1 inch long, dry and spiny.

This cactus is common in the East Mojave. You'll find it on the hillsides between Government Holes and Rock Springs. It also grows at Hole-in-the-Wall.

GRIZZLY BEAR CACTUS (*Opuntia erinacea var. ursina*) is rarer in the Scenic Area. It has more and longer spines—up to 6 inches long.

Old Man Cactus in bloom

PANCAKE CACTUS—PRICKLY PEAR
Opuntia chlorotica
Flower: Yellow

Cactus Family - *Cactaceae*

Pancake Cactus is erect and treelike growing to 6 or 7 feet tall in maturity. It has a definite trunk, woody-looking on older plants, with dark spines near the base. It is much-branched with flat round pads up to 7 inches wide. The pads are covered with glochids and spines growing from areoles placed in symmetrical rows across the pad. The yellow spines, up to an inch long but of unequal sizes, generally point downward on the pad, except those growing right on the rim.

Blooms, which occur in summer, are yellow, about 2 inches wide. Fruit is purple and fleshy, filled with tiny black seeds. The spineless fruit is a tasty tidbit for desert birds and animals. Many animals also eat the pads.

Birds use this cactus for a protected nest site.

Good specimens of Pancake Cactus can be found above Rock Springs.

Pancake Cactus and small Barrel Cactus

PENCIL CHOLLA—DIAMOND CHOLLA—DARNING NEEDLE CACTUS
Opuntia ramosissima
Flower: Green, Yellow, Straw, Brown

Cactus Family - *Cactaceae*

From a distance, when its long needles catch the sun's light, this cactus sometimes resembles a shrub in bloom. It's a formidable well-armed cactus, however, growing to 5 feet tall with many compact branches. Its Latin name, ramosissima, means "very much branched" and is also given to Blackbrush (*Coleogyne ramosissima*)—that compact, twiggy, ubiquitous plant of the high desert.

The joints of this cactus have diamond-shaped patterns out of which grow the yellow-sheathed, 2-inch-long spines. Oddly, some of the slender, pencil-sized stems do not have any spines at all.

The small flowers are inconspicuous but the spiny, woolly fruit is quite noticeable. Bloom is in late spring or summer.

Pencil Cholla is common in the East Mojave. Look for it along Cedar Canyon Road near Kelso-Cima Road, along Kelbaker Road, and near Hole-in-the-Wall.

PINCUSHION CACTUS
Coryphantha vivipara deserti
Flower: Pink

Cactus Family - *Cactaceae*

You're likely to just accidentally come across this little cactus. They are small and blend into their surroundings so, unless the delicate pink flowers are open in May or June, they often go unnoticed. The plants have 1 to 4 or 5 chubby stems, about 4 inches tall. The stems have many deep nipple-like tubercles. Straight spines grow from areoles on the tubercles. Flowers have many slender petals.

The very young, small cacti, just poking up an inch or so above the ground, are often dug up and destroyed by small animals. I've resorted to putting wire cages around several on our property.

Pincushion Cacti are fairly common in a limited elevation zone—between 4,000 and 6,000 feet—in the East Mojave. They prefer soils rich in limestone. Look for these interesting plants on the Teutonia Peak Trail Walk and near the Providence and New York Mountains.

Another small cactus you may see in the lower elevations, below 4,000 feet, is **FISHHOOK** or **NIPPLE CACTUS** (*Mammillaria tetrancistra*). It's just a few inches high but when in bloom quite noticeable and charming. The delicate petals have rose-colored stripes. The central spines are hooked. A good spot to look for these plants is on the Volcanic Rock Hill Walk.

Pincushion Cactus with Rattlesnake Weed

SILVER CHOLLA—GOLDEN CHOLLA
Opuntia echinocarpa var. echinocarpa
Flower: Green, Yellow

Cactus Family - *Cactaceae*

   This cactus is generally stouter and shorter than its close look-alike, Buckhorn Cholla. Along Cedar Canyon Road I've noticed Silver Cholla growing at the lower elevations at both ends of the road and Buckhorn Cholla growing in the higher mid-section of the road. Their habitats do overlap, however, and they may even hybridize.
   Silver Cholla grows from 2 to 5 feet high with many cylindrical branches. Joints are 4 to 8 inches long, covered with silver or golden spines about 1 inch long. Each spine is covered with a removable papery sheath.
   Flowers are an interesting greenish-yellow and bloom in May and June. Fruit is densely spiny.
   This cactus is common in the Scenic Area.

TEDDY BEAR CHOLLA—JUMPING CHOLLA
Opuntia bigelovii var. bigelovii
Flower: Yellow

Cactus Family - *Cactaceae*

John Milton Bigelow was part of the Pacific Railroad Survey of 1853-4 and collected most of the survey's plants. His most famous plant discovery was this cholla that bears his name.

The Jumping Cholla doesn't really "jump" but, as careful as you try to be, a joint will attach itself to your clothes or pant legs when you walk through a thicket of these angelic-looking devils.

This species is easy to recognize by its treelike shape, dense stubby upper branches, and dingy, old, black trunk. These cacti are generally 3 or 4 feet high but can grow over 6 feet. The joints are densely spine-covered. These 1-inch-long spines are barbed and difficult to remove. Joints fall off easily and the surrounding land is covered with them. Don't walk your dog near chollas.

The yellow flowers, which bloom in April, are fairly inconspicuous and seeds are often sterile. Propagation is from the fallen joints. The fruit is covered with glochids.

Wood rats use the joints to protect their dwellings. Birds nest in the dense branches.

The best place to view Teddy Bear Cholla in the East Mojave is just outside the Scenic Area while driving from Needles west toward Barstow on Interstate 40 in the Sacramento Mountains. There is an area set aside there for the protection of these plants—the Bigelow Cactus Gardens.

BLUE GRAMA GRASS
Bouteloua gracilis

Grass Family - *Poaceae*

Besides being among our most graceful plants, grasses aid in the control of soil erosion and provide valuable forage for native animals and cattle. This perennial grass is densely tufted, growing 12 to 18 inches tall. It tolerates cold, drought, and poor soil. The seed grows on thin spikes in 2 rows. When dry these spikes curl downward and in, sometimes forming a circle.

The older plants form "fairy circles" by growing outward in a ring and dying in the center.

Blue Grama grows in the high elevation Piñon-Juniper country of Mid-Hills, Clark and New York Mountains, and on the hillsides along Cedar Canyon Road near Government Holes.

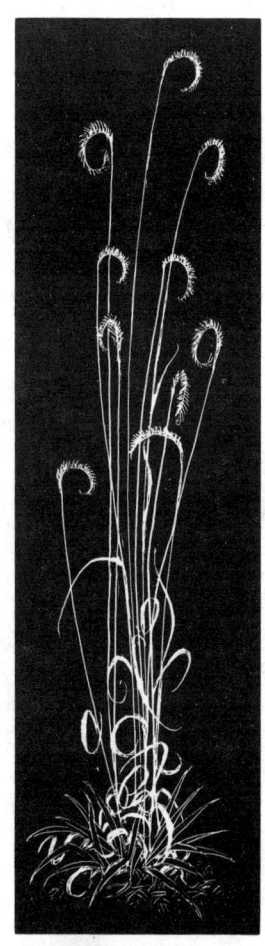

DESERT STIPA—DESERT NEEDLE-GRASS
Stipa speciosa

Grass Family - *Poaceae*

Desert Stipa, a densely tufted perennial bunch grass from widespread roots, is another native species. The little bracts surrounding the inconspicuous flowers are hairy. The awn (the bristle-like appendage that sticks up above the bracts) is hairy at the base. The upper portion of the awn is bent.

A friend asked me, "What's that grass that looks like miniature Pampas Grass?"—an apt description.

Look for Desert Stipa in the Rock Springs-Government Holes area.

GALLETA GRASS
Hilaria jamesii

Grass Family - *Poaceae*

Galleta is a hardy, stiff, perennial grass growing from underground rhizomes. It can grow to almost 2 feet tall. Stems are erect and slender. The flowering spikes, at the tops of stems, are about 2 or 3 inches long. Each spikelet (flower cluster) is pressed flat to the blade, in groups of 3. Spikelets are hairy, especially at the base. They are green and straw-colored, tinged with purple.

Galleta Grass is common in Piñon-Juniper country around Mid-Hills, Clark Mountain and Pinto Valley.

The rugged grass growing in large, dense clumps on the lower portions of Kelso Dunes is **BIG GALLETA** (*Hilaria rigida*). The woolly stems are rigid, up to 2 1/2 feet tall. The spikelets (flowering portions) are in clusters of 3. After the seeds have fallen off you will note that the tops of the stems on both these grasses are wavy. They are desert natives and are heavily grazed.

Big Galleta is more common than Galleta Grass throughout the East Mojave. It will generally be in sandy areas but also occurs in Blackbrush Scrub.

Galleta Grass

INDIAN RICEGRASS—SAND BUNCHGRASS
Oryzopsis hymenoides

Grass Family - *Poaceae*

Grasses of the East Mojave contribute much to the beauty of the landscape. Indian Ricegrass is one of the most lovely, delicate grasses in this area. If not eaten, it remains standing, dried and straw-colored, still with a few seeds, until new growth pushes up in spring from perennial roots.

This grass is erect with green leaves bent on the midrib, almost folded in half. The upper branches are delicately divided with seeds at the tip ends. The plant grows to about 18 inches high.

Indian Ricegrass is eaten by many animals, including cattle. Native Americans gathered the small seeds for food.

This grass grows on Kelso Dunes and throughout the Scenic Area. It has one of the greatest elevational gradients of a grass in the East Mojave. It's common but scattered in Cedar Canyon and Black Canyon.

Delicate upper portion of Indian Ricegrass

ANDERSON LYCIUM—DESERT THORN—WOLFBERRY
Lycium andersonii
Flower: White, Lavender

Nightshade Family - *Solanaceae*

This is a shrub that looks like it's dead all winter.

The lower wood is very dark. There are sharp spines on the stems. The leaves are about 1/2 inch long, fleshy and succulent. Flowers grow in the leaf axils. The small greenish-white to lavender blooms are funnel-shaped with 4 to 6 lobes at the top. Bloom begins in March at the lower elevations. Lycium can grow to 5 or 6 feet high and is heavily grazed in this open range land. Local cattlemen call it June Berry.

The round fruit is about 1/4 inch long and bright red to orange. This little berry is quite tasty, if somewhat astringent. Native Americans ate them and they are a favorite of many birds and animals.

Lycium shrubs are common and can be seen at the intersection of Black Canyon Road and Cedar Canyon Road, and on the volcanic hills near Kelbaker Road in the Cinder Cone area, and along Cima Road.

APACHE PLUME
Fallugia paradoxa
Flower: White

Rose Family - *Rosaceae*

Delicate white blossoms and soft feathery-tailed seed clusters decorate this otherwise somewhat straggly, unkempt-looking shrub.

Leaves grow alternately on the erect, flexible branches. They are rarely over 1/2 inch long and are lobed at the top into 3 to 5 divisions. Solitary flowers bloom at the top of leafless stems (peduncles). The white blooms are about 1 inch wide with 5 petals, 5 sepals, many stamens and pistils. The styles mature into long pinkish-white plumes. Flowers and plumes may be on the bush at the same time. The plumes often remain on the plant all year but get matted and beaten down by rain and snow.

It's a real "photo opportunity" to find a large stand of these shrubs in plume, backlit by the sun.

Some plant books list Apache Plume as being deciduous, but I find the many specimens here at Quail Rock to have leaves all year—albeit in winter they have dropped quite a few.

The "paradox" of the Latin name is either, 1. that the plant's discoverer found, after careful examination, it was not what he first thought it to be or, 2. plumes and blooms are together at the same time. Authorities seem to differ on this subject.

Apache Plume grows above 3,500 feet elevation to about 6,000 feet beside washes, near boulders and on slopes. It blooms in late April through June and again, depending on summer rains, in August and September. The late summer bloom is not as heavy.

There are several large shrubs beside the sandy track off Cedar Canyon Road leading to Camp Rock Springs, many in Watson Wash where it crosses Cedar Canyon Road below Rock Springs, and scattered throughout Caruthers Canyon.

ithery plume and 5-petaled flowers of Apache Plume

BITTERBUSH—ANTELOPE or BUCK BRUSH—SWEET SAGE
Purshia glandulosa
Flower: White

Rose Family - *Rosaceae*

Bitter is right! If you're thinking of chewing a Bitterbush leaf, be sure to have something handy to counteract the taste. The local name of Sweet Sage was given to me by a rancher whose cattle seem to enjoy the flavor. The plant is neither sweet (except to a cow) nor a sage.

Bitterbush is an evergreen shrub, up to 8 feet tall, growing above 3,000 feet. The waxy leaves glisten in sunlight. Close examination reveals tiny, white, gland-dots on the upper leaf surface. The 1/4 inch long leaves grow in crowded clusters. Each leaf is 3-cleft with slightly rolled-under margins. They are strongly scented.

This shrub is a close relative of Cliff Rose and at a casual glance it's difficult to tell them apart when there are no flowers or fruit. The leaves of Cliff Rose are 3 to 5-cleft.

The cream-colored flowers of Bitterbush are smaller and less showy than Cliff Rose but beautiful all the same, covering the whole bush. They are small, sweet-smelling, with 5 sepals and 5 petals. Bloom begins in April.

The interesting, dry, 1-seeded fruit is up to 1/2 inch long, triangular shaped, tapering to a narrow tip with one tail (hair) growing out of it. This hair is a persistent style (the part of the pistil that connects stigma and ovary).

Bitterbush is common in the higher elevations such as the Mid-Hills area. You'll see it on the Teutonia Peak Trail and in Caruthers Canyon.

CALIFORNIA BUCKWHEAT, DESERT BUCKWHEAT
Eriogonum fasciculatum ssp. polifolium
Flower: White, Pink

Buckwheat Family - *Polygonaceae*

Eriogonum is a huge genus with over 150 species. *Eriogonum fasciculatum* itself has several subspecies. It's a challenge to identify many of the buckwheats but this one is common and easily learned.

In Latin, "fascis" means "bundle" and that is how the leaves grow on California Buckwheat—in tiny bundles. The shrubs are usually low-growing, about 2 feet high. Leaves are gray-green, 1/4 inch long, and rolled to the underside. Flowers, without petals, form in dense heads at the top of leafless stems from April to June. The whitish-pink color comes from the calyx.

The bushes are attractive in fall and winter since they retain the dried flower heads which turn a lovely russet red.

California Buckwheat lines many of our desert roadsides. Look for it along Cedar Canyon Road.

In August, when California Buckwheat is starting to turn color, another member of this family, **WRIGHT BUCKWHEAT** (*Eriogonum wrightii ssp. wrightii*), begins blooming. Crowded, narrow leaves, woolly-white on the underside, grow on the lower third of the plant. There are many flowering stems which are 2 or 3 times branched. Tiny, delicate blooms are scattered along the branches. This is a low-growing sub-shrub that looks lacy when in bloom while California Buckwheat appears to have large pieces of popcorn at the top of its stems. You can find this plant at Camp Rock Springs.

A single flowering stem of California Buckwheat

COOPER LYCIUM—PEACH THORN—BOXTHORN
Lycium cooperi
Flower: White

Nightshade Family - *Solanaceae*

This Lycium is similar to Anderson Lycium with several variations. The leaves of Cooper Lycium grow in little bundles, are longer and not as fleshy. White funnel-shaped, 1/2-inch-long flowers have 5 distinct lobes and look somewhat like a star. They're streaked with green and lavender veins.

The fruit is green and constricted at the top as if someone had tied a tight string around it. The fruit of this shrub is unpalatable. A local name is Mock June Berry.

Cooper Lycium is common in the East Mojave, often growing in association with Anderson Lycium. Good specimens grow near Cima Road.

DESERT ALMOND
Prunus fasciculata
Flower: White

Rose Family - *Rosaceae*

The fruit of this large deciduous shrub is much more noticeable than the tiny, white, cupshaped blooms.

The intricately-branched framework of grayish bark makes Desert Almond easily recognizable, even in winter.

New leaves begin appearing in March, closely followed by the bloom. At the same time, many shrubs become covered with silky tent-caterpillar webs. The main activities of these caterpillars seem to be communal sunbathing and roof repairs.

Small green leaves grow in crowded bunches (fascicles) on short branchlets. Flowers are about 1/8 inch long with 5 sepals and 5 petals. Fruit is an oval drupe (a fleshy fruit containing a stone with a kernel, like a peach) up to 1/2 inch long. These are favored by antelope ground squirrels.

In summer, the husks of the fruit litter the ground beneath the bush where the squirrels have dined. The fruit has a delicious almond flavor. To sample, peel off the fuzzy outer coat, then crack the shell to reach the kernel.

Desert Almond is a common shrub above 2,500 feet elevation. It lines the wash along lower Wildhorse Canyon Road and fills the entrance of Caruthers Canyon.

Tiny flowers of Desert Almond

Fruit of Desert Almond

DESERT ALYSSIUM
Lepidium fremontii
Flower: White

Mustard Family - *Brassicaceae*

This is one of the earliest blooming plants in the East Mojave, often decorating the lower elevation roadsides in late February. It is a low, rounded, much-branched perennial subshrub, growing to a height of 1 or 2 feet.

Leaves are linear, less than 2 inches long. Masses of tiny, white, sweet-smelling flowers are produced, each with 4 petals 1/8 inch long. The fruit is a tiny pod, slightly heart-shaped.

Like many other members of the Mustard Family, the leaves are tangily pungent.

You'll find it in bloom near Caruthers Canyon in late spring.

DESERT HOLLY
Atriplex hymenelytra
Fruit: Light green

Pigweed Family - *Chenopodiaceae*

Powdery white, hollylike leaves on this low rounded bush are a stark contrast to the black volcanic rock on which it grows in the Cinder Cone area off Kelbaker Road.

The plant is 1 to 3 feet tall, with many white branches. The silvery round leaves—1/2 inch to 1 inch long—have deep but irregular indentations. They feel like fine suede.

The inconspicuous, pinkish male and female flowers grow on different plants. In early spring the female plants are covered with green fruiting bracts.

A prime location to see Desert Holly is on Kelbaker Road, about 13 miles east of Baker where a black lava hillside comes right down to the north side of the road. Walk behind the hill and you'll find Desert Holly growing on the back slope and in the rocky desert to the north. Also in this spot are good specimens of Cottontop Cactus, Hedgehog Cactus, Beavertail Cactus, Anderson Lycium, Desert Willow and Desert Trumpet. (See Volcanic Rock Hill Walk.)

FLAT-TOP—PLUME BUCKWHEAT
Eriogonum plumatella
Flower: White

Buckwheat Family - *Polygonaceae*

Thin, erect, closely-packed stems make this member of the large Eriogonum genus look as if it has been bundled up with string and just freshly transplanted from the nursery. The plant "fluffs out" during late spring and summer growth.

Stems on this perennial plant are usually about 2 feet tall, covered with soft white hairs and forked in the upper portions to form flat crowns. Half-inch, oblong leaves, on the lower portions of stems, are rolled to the underside. Tiny flowers grow close together on the forked branchlets. Their white color is from the calyx. Bloom is in late summer.

Many Flat-Tops are along the roadsides of Kelbaker Road near the Granite Mountains and Kelso-Cima Road near Cedar Canyon Road.

SANDPAPER PLANT
Petalonyx thurberi ssp. thurberi
Flower: White

Loasa Family - *Loasaceae*

Go ahead—touch this plant. The stems and leaves feel just like fine sandpaper. The looks are deceiving because the hairs that cause this rough sensation are very short and pressed flat against the stem.

The plant has many stems growing to about 2 feet tall. Alternate leaves are attached directly to the stems. Tiny white, fragrant flowers bloom on short terminal spikes. Bloom begins in late May and continues through summer. In winter the stems turn white but you'll still be able to identify the plant—by touch.

The cocklebur gave the inventor of Velcro his idea for that product but this family would have given much better examples.

The base of Kelso Dunes has a large population of Sandpaper Plants. You'll also find it along Kelbaker Road and Black Canyon Road near Essex Road.

SPINY MENODORA—TWIN FRUIT
Menodora spinescens
Flower: White

Olive Family - *Oleaceae*

If you go on the suggested hike in Cedar Canyon Wash where it crosses the main road 3.9 miles east of Kelso-Cima Road, you'll see many specimens of Spiny Menodora. Teutonia Peak Trail is also a good place to find them.

This is a low shrub, growing to about 3 feet tall, spreading 3 or 4 feet, with short, spiny, yellow-green interlaced stems. Leaves are small, linear and grow alternately on the stems. Flowers are funnel-shaped, 1/4 inch long, white, sometimes with a purple cast. There are only 2 stamens. Bloom is in April and May. The fruit is a purple-tinged 2-celled capsule, with 2 seeds in each cell.

Spiny Menodora is often a "nurse" plant for young Joshua Tree seedlings. If a Joshua Tree seed drops on the open ground, it does not have much of a chance of growing to maturity. But its chances are greatly improved if it takes root and is nurtured under the spiny branches of a shrub until it is big enough to withstand the trials of desert life on its own.

SQUAW WATERWEED—DESERT BACCHARIS—SARATOGA
Baccharis sergiloides
Flower: White

Sunflower Family - *Asteraceae*

A local old-timer told me that in homestead days, when choosing a site to drill a well, the two plants that were true water indicators were Saratoga and Hackberry. That was the only time I've heard Squaw Waterweed referred to as Saratoga.

It is usually leafless but the 3 to 6-foot-high broomlike stems are always green. The bright green leaves of spring are somewhat oval. The very small whitish male and female flowers bloom in compound clusters, called panicles, during late spring and summer. Dried flower heads remain on the stems all year.

This plant grows in the wash above Rock Springs, along the streambed in Caruthers Canyon, and in many of the washes in the Scenic Area.

Along Piute Creek you will find another member of this genus, **WATER-WALLY** or **SEEP WILLOW** (*Baccharis glutinosa*). It is a willowlike shrub with bright green linear leaves, 2 to 5 inches long, on stems up to 9 feet tall.

A small Squaw Waterweed

WINTER FAT—WHITE SAGE
Ceratoides lanatum
Flower: White

Pigweed Family - *Chenopodiaceae*

White, woolly, cotton tufts are this small shrub's distinguishing feature. The plant grows from 1 to 3 feet high and has white, hairy, flexible stems with gray-green, 2-inch-long, linear leaves with rolled-under margins. Tiny male and female flowers grow on the same bush. The female flowers have long silky hairs. Fruits are densely covered with cottony tufts. Bloom is from March to June but the cotton-covered stems are conspicuous throughout fall and winter. The stems make attractive dried-flower arrangements, lasting several years.

It is a valuable forage plant, especially in winter when other shrubs have shed their leaves, hence the common name, Winter Fat. Lanatum in Latin means "woolly."

Winter Fat grows above 2,000 feet elevation in the Scenic Area. There are good specimens along Cedar Canyon Road about 5 miles west of Lanfair-Ivanpah Road.

"Cotton" tufts on a flowering stem of Winter Fat

WOOLLY BRICKELLIA
Brickellia incana
Flower: White

Sunflower Family - *Asteraceae*

You'll see this small gray shrub lining many of the roadsides in the East Mojave. Kelbaker Road and lower Black Canyon Road have good specimens.

White woolly, oval leaves are about 1 inch long, without stems. The disk flower heads, almost 1 inch long, are surrounded by a series of long, woolly, purple-tinted bracts. Showing above the bracts are white pappus bristles topped by maroon stamens and pistils. The whole effect is softly pleasing.

Woolly Brickellia is generally 2 to 3 feet tall. Bloom is from April to October.

Another Brickellia you may meet in the higher elevations of the Scenic Area is **PIÑON BRICKELLIA** (*Brickellia oblongifolia var. linifolia*). It has several stems, up to 15 inches tall, branching from a woody base. Leaves are linear, less than 1 inch long, without petioles. The white disk flower heads are narrow, about 1/2 inch long, with long involucre bracts. Piñon Brickellia is not too common but can be found in the higher elevations near rocky washes and in stony soil.

Woolly Brickellia

Flowering stem of Piñon Brickellia

BARBERRY—HOLLY-GRAPE—MAHONIA
Mahonia (Berberis) haematocarpa
Flower: Yellow

Barberry Family - *Berberidaceae*

Barberry is an evergreen, spiny, large shrub with holly-type leaves. Masses of small yellow flowers cover the plant in May and June. The entrance road to Caruthers Canyon is lined with Barberry. There are a few scattered plants near Cedar Canyon Road by Black Canyon and on the slopes above Rock Springs.

Flowers are followed by grapelike juicy berries said to make delicious jellies and jams.

Bark and roots are a source of yellow dye. It is one of the prized medicinal plants of the area due to the yellow alkaloid, berberine, and is used for laxatives, antibacterial washes, fevers, and hangovers. Tea made from the root is quite bitter.

Barberry is a beautiful landscape ornamental.

BIG SAGEBRUSH—GREAT BASIN SAGEBRUSH
Artemisia tridentata ssp. tridentata
Flower: Yellow

Sunflower Family - *Asteraceae*

Sagebrush is one of the most common plants of the West and is an indicator plant of the Great Basin Desert, as is the Joshua Tree for the Mojave Desert. It is found only at the higher elevations in the East Mojave, and covers many acres above 4,000 feet. The silvery aqua-green shrubs, 2 or 3 feet tall, along Cedar Canyon Road near Black Canyon, are Sagebrush.

It is not a "sage" at all but, like true sages, is very aromatic when crushed or after a rainstorm. The wood smells delicious when burned. I find the leaves bitter to the taste but rabbits nibble them with relish.

The Latin name "tridentata" refers to the leaf tip with 3 teeth on the upper edge.

Sagebrush is an evergreen shrub—or evergray—that changes little during the year. The small yellow flower spikes bloom late, usually in October, and remain dried on the top of the plant all year, looking much the same as when in bloom.

Plants may be 50 years old. Many of the ancient trunks are twisted and gnarled. One specimen in Cedar Canyon Wash is over 8 feet tall.

Carl often paints Sagebrush. He finds the rugged trunks, the fallen dead wood at the base of the plants, and the convoluted forms to be artistically pleasing.

Lichen sometimes grows on the base of the old plants.

Big Sagebrush

BLACK-BANDED RABBITBRUSH—STICKY RABBITBRUSH
Chrysothamnus paniculatus
Flower: Yellow

Sunflower Family - *Asteraceae*

This showy Rabbitbrush is so named because a smut fungus causes black bands on the stems. The bands are usually small, no more than 1/2 inch wide, and don't seem to harm the plant.

The bright green, resinous leaves are needle-sized, about 1 inch long, on stems up to 7 feet tall.

These shrubs usually begin blooming in September and are covered with hanging flower clusters. Like Golden Rabbitbrush, each rayless flower is very small but they bloom in such profusion they almost conceal the stems. Hundreds of butterflies flit about the flowers.

This plant fills the wash that crosses Cedar Canyon Road 3.9 miles east of Kelso Cima Road and can be found in almost all of the washes in the East Mojave.

BLACKBRUSH
Coleogyne ramosissima
Flower: Yellow

Rose Family - *Rosaceae*

Blackbrush takes over after Creosote Bush and Joshua Tree leave off, marking the upper division of the Lower Sonoran Zone.

It is well named. The wood is very dark and appears quite black after a rain.

The pronunciation of Coleogyne had me stumped for awhile. It's Ko-lee-ah'-jih-nee. That's Greek for "sheathed ovary." Ramosissima means "much branched" in Latin. This is a very compact, stiff, twiggy bush, usually not more than 2 or 3 feet tall. It grows very equally spaced in large stands.

The tiny green leaves grow in opposite bundles and are deciduous. The solitary flowers do not have petals. The yellow color is from 4 small sepals. Each flower has 30 or 40 stamens. The blooming period is from April to July.

Blackbrush Scrub is a prevalent plant community in the East Mojave, especially on north slopes of the New York, Ivanpah and Clark Mountains.

Large stands of Blackbrush grow along Black Canyon Road near Hole-in-the-Wall.

A flowering branch of Blackbrush

BLADDERPOD
Cleome isomeris (Isomeris arborea)
Flower: Yellow

Caper Family - *Capparidaceae*

It's hard to find a time when one of these dependable shrubs is NOT blooming. In the middle of winter, their cheery yellow flowers do much to lift the spirit when other plants seem to be just barely hanging on. It is very common along Kelbaker Road, at elevations below 4,000 feet. Look for it near I-40, and on the other end of Kelbaker Road, about 10 miles east of Baker.

Bladderpod is a much-branched, graceful shrub (albeit smelly), up to 8 feet high. The leaves are alternate on the stems; each one has 3 leaflets.

Yellow flowers grow in terminal clusters. They are showy, with 4 sepals, 4 petals, each 1/2 inch long, and 6 stamens that protrude above the corolla.

Fruit becomes a plump 2 inch pod which dangles down on a long stem. Flowers and fruit may be on the plant at the same time.

BURROBUSH—WHITE BURSAGE—BURROWEED
Ambrosia dumosa
Flower: Yellow

Sunflower Family - *Asteraceae*

This rather insignificant looking, low, compact, twiggy bush is one of the more dominant plants in the West, often growing as a co-dominant with Creosote Bush. It is 1 or 2 feet high and scarcely calls attention to itself. The 1/2-inch-long green leaves of spring turn gray and fall off early. Separate male and female flowers are on the same bush, the female heads becoming, in fruit, a beaked bur with 20 to 30 spines. The entire fruit is no more than 1/4 inch long.

Burrobush flowers in spring and sometimes again in fall. The plant is heavily browsed.

Look for the little gray understudy growing alongside the larger Creosote Bushes in the lower elevations—below 4,000 feet—of the Scenic Area and throughout the deserts of the West.

A single flowering stem of Burrobush

CALIFORNIA BRICKELLIA—BRICKELL BUSH
Brickellia californica
Flower: Yellowish-white

Sunflower Family - *Asteraceae*

Leaves on this graceful perennial shrub are alternate on the stems. They are almost heartshaped, about 1 inch long, deeply veined, with scalloped-toothed margins. Flower heads, in clusters at the ends of short branches, consist of disk flowers only. The involucral bracts turn papery and some remain on the bush, along with a few dried-out leaves, all year, thus making identification of this plant fairly easy, even in winter. California Brickellia has many white stems growing to about 4 feet high.

It's late to leaf out in spring. Bloom is in late summer through early fall.

Look for California Brickellia growing among the boulders lining the wash just below Rock Springs.

Leaves of a small California Brickellia

CALIFORNIA CROTON
Croton californicus var. mohavensis
Flower: Yellow

Spurge Family - Euphorbiaceae

Large sand hummocks are formed by California Croton on the lower portions of Kelso Dunes. It's a low, spreading plant with many branches. The light olive-green leaves, less than 1 inch long, are narrowly oblong, on petioles about 1/2 inch long. They were used as poultices by Native Americans. Tiny yellow male and female flowers are on separate plants. Bloom begins in late spring.

The stems whiten and wither in winter and the wind disperses some of the Croton hummocks, but these plants do much to stabilize the dunes. Croton, in a more erect form, can also be found along sandy roadsides or in washes.

CHEESEBUSH—BURROBUSH
Hymenoclea salsola
Flower: Yellow

Sunflower Family - *Asteraceae*

There's a good reason to learn the Latin botanical name for Cheesebush. Plant books seem to be equally divided in calling this plant Cheesebush (that's what I've always called it) or Burrobush (the name I use for *Ambrosia dumosa*). Hymenoclea means "enclosed membrane"—hymen is Greek for "membrane," and kleio means "to enclose." Salsola is the botanical genus name for Tumbleweed, which it's supposed to resemble. My opinion is that it doesn't resemble that noxious weed at all—I'm quite partial to Cheesebush, the common name we'll use here, which refers to the odor of the crushed leaves. Quite pleasant, I think.

Once you learn to recognize Cheesebush, you'll be surprised at how common this plant is. It lines many of the roadsides in the East Mojave. This shrub appears to colonize disturbed areas both natural and man-made such as washes and road embankments, while Tumbleweed colonizes only man-made disturbances such as road cuts or ORV areas.

The deciduous shrub is 3 or 4 feet high. Most of the year it's a straw-colored, leafless bunch of stems. In spring, the very narrow threadlike leaves sprout, and suddenly great portions of the desert seem to come to life.

Flowers are so inconspicuous they're hardly seen. Male and female flowers are on the same plant. The fruit of the female flower is highly noticeable. The small fruit is surrounded by 7 to 12 papery silver wings. Sometimes the wings have a purple cast. It is during this stage the plant is most attractive. Local ranchers call this White Berry Bush. Their cattle eat only the fruit.

CLIFF ROSE—QUININE BUSH
Cowania mexicana var. stansburiana
Flower: Yellow

Rose Family - *Rosaceae*

Cliff Rose is one of the loveliest plants of the high elevations in the East Mojave. It is evergreen, sometimes growing to the size of a small tree, with rugged shredding bark. The trunks are often picturesquely twisted.

Before bloom it is difficult to distinguish this plant from Bitterbush (*Purshia glandulosa*) without careful examination of the leaves. The gland-dotted leaves of Cliff Rose are 3 to 5-cleft. In bloom, the 5-petaled, sweet-smelling, creamy yellow flowers of this bush almost cover the plant. Long feathery plumes develop in the fruiting stage. Bloom is from April to July.

Cattle browse extensively on the shrub. The soft bark was used by Native Americans to make clothing, sandals and rope. They also used it to pad cradleboards, thus inventing the first biodegradable disposal diapers.

You'll find Cliff Rose in the higher elevations—above 4,000 feet—in the mountainous areas of the East Mojave. The canyons leading up Pinto Mountain have excellent specimens of this plant.

CREOSOTE BUSH—GREASEWOOD
Larrea tridentata (divaracata)
Flower: Yellow

Caltrop Family - *Zygophyllaceae*

If you're a desert lover you are no doubt already acquainted with Creosote. It is the Number One most successful and widespread plant of the deserts. It covers vast stretches of land over much of the East Mojave, often with little gray Burrobush in its shadow. This plant community covers 80% of the California deserts.

Creosote grows properly spaced as if planted by a fussy gardener. All desert plants compete for water and some, like Creosote, may exude a substance in their roots to repel the growth of other plants.

If someone could bottle the aroma of Creosote after a rain, I'd be their best customer. That aroma is the true and familiar scent of the desert.

Creosote Bush is a graceful, airy shrub with many long flexible stems. It's usually 2 to 4 feet tall but along roadsides and washes where there is more available moisture, it can attain heights of 10 or 12 feet.

The shiny, oily, resinous leaves are placed opposite on the branches. They are made up of 2 leaflets, joined at the base, not more than 1/2 inch long.

The yellow flowers have 5 sepals and 5 petals, slightly twisted. Fruit is a little round capsule covered with woolly hair.

Did it just snow on the desert? No—the hundreds of Creosote Bushes are covered with their white fuzzy seed pods. The fruit ripens and separates into five parts, each with one seed. Creosote seeds are difficult to germinate.

Bloom is usually about March through May but elevation and rainfall cause variations.

A gall midge (a fly) causes the large round swellings so numerous on the plants. Another insect, a scale, deposits orangish-red lac on the stems, a resinous substance which was used by Native Americans to waterproof their baskets.

Creosote tea is a popular concoction for various ailments (about one sip is my limit—I find it most unpalatable) and is used medicinally both internally and externally.

Creosote marks the division between the upper and lower division of the Lower Sonoran Life Zone. You will note this mark, which occurs rather abruptly, on Cedar Canyon Road heading east. Near the corral close to the end of the pavement is also the end of Creosote Bush. You then pass through a transition zone where Blackbrush is dominant and soon enter a Sagebrush Scrub plant community. You re-enter Creosote Scrub as you drop in elevation heading east.

Creosote Bush

DESERT OLIVE—TANGLEBRUSH—HACKBERRY
Forestiera neomexicana
Flower: Yellow

Olive Family - *Oleaceae*

Desert Olive is not too common in the Scenic Area but you will probably come across it in a walk above Rock Springs, in Caruthers Canyon, or other spots with high water tables. It, and Squaw Waterweed, are the "sure fire" water indicator plants the old-timers used when drilling wells.

Hackberry, strictly a local name for this plant, is not related to the true Hackberry (*Celtis sp.*), however, both are good water indicators.

Desert Olive is a deciduous, twiggy, crookedly-branched small tree or large shrub with gray bark, growing to about 10 feet tall. Leaves are 1 or 2 inches long with slightly toothed margins.

Blooms are very tiny with no petals. The stamens give the flowers a yellow cast. The fruit is a small, dark, olive-shaped drupe (a 1-seeded fruit containing a pit). Bloom is in April and May.

Small fruit of Desert Olive

DESERT SENNA—DESERT CASSIA
Cassia armata
Flower: Yellow

Pea Family - *Fabaceae*

The first time I saw Desert Senna in bloom on Kelbaker Road about 10 miles from Baker, in April, I jumped from the car to take a picture and was engulfed by a wonderful sweet scent. Most of the year it appears quite dead, but following a good rainy season the low round bushes burst into masses of yellow blossoms.

There are few leaves and these drop off after the bloom. The corolla is made up of 5 petals. Flowers are followed by a spongy thin pod.

It is locally called Scotch Broom which is the common name usually given *Cytisus scoparius*, another plant of the Pea Family with pea-like flowers.

Desert Senna is common near sandy washes. The highest limit of its range is about 4,000 feet. Look for it along Black Canyon Road near Essex Road, near Kelso Dunes and near Kelso-Cima Road.

FOUR-WING SALTBUSH—WINGSCALE
Atriplex canescens ssp. canescens
Flower: Yellow

Pigweed Family - *Chenopodiaceae*

This striking shrub is most noticeable in late summer and fall when the female plants are covered with membranous fruits, shaped in a four-winged arrangement like a butterfly on a mirror, surrounding a small seed.

The plant is dioecious—male and female flowers are on different plants. Leaves are gray and narrow, turned up into a little trough. Cattle find these leaves to be on the top of their list of epicurean delights. Locally this plant is called Chamise, a common name used for several other shrubs in California.

Saltbush is related to Tumbleweed; also in the same family are beets and spinach.

The plant is usually 3 to 5 feet tall, densely branched, and keeps most of its leaves year round. New leaves appear in April and the flowering yellow spikes bloom in June.

It is widespread throughout the East Mojave. You can find it on the Cedar Canyon Wash Walk. The female plants usually keep some of their seed pods all year which makes identification easier.

Another common saltbush in the East Mojave is **ALLSCALE** or **CATTLE SPINACH** (*Atriplex polycarpa*) which you can find in alkaline soils such as near the Cinder Cones. It is intricately branched, about 5 feet tall, with crowded gray leaves. The small fruiting bracts are toothed. Bloom is in late summer.

Fruiting stage of a female Four-Wing Saltbush

GOLDENEYE—DESERT SUNFLOWER
Viguiera deltoidea var. parishii
Flower: Yellow

Sunflower Family - *Asteraceae*

Goldeneye is a small shrub, 1 to 3 feet tall, with a woody base. The lower leaves on the stem grow opposite. They are deep green, rough, hairy, toothed and slightly triangular-shaped. Flowers grow atop long leafless stems. Each has 8 or more yellow rays. Bloom is in early spring and again in early fall.

Goldeneye, locally called Wild Aster, prefers mid to high elevations and can be found in the Providence and New York Mountains. There are many good specimens on Black Canyon Road south of Hole-in-the-Wall, and near Bonanza King Mine.

GOLDENHEAD
Acamptopappus sphaerocephalus
Flower: Yellow

Sunflower Family - *Asteraceae*

The common name for this charming little shrub is appropriate. A clear, light yellow glow seems to radiate from the many small flowers. White-barked stems are usually about 2 feet tall. Small green, alternate leaves are linear, with a tiny spine tip. You may need a hand lens to see the sharp point. Goldenhead has only disk flowers. Bloom is from April to June. It is common in most of the East Mojave. Ft. Piute, Lanfair Valley, Cedar Canyon and Black Canyon near Essex Road are good spots to find it.

Acamptopappus means "stiff pappus" (the bristles at the tops of the seeds) and sphaerocephalus means "sphere-headed."

GOLDEN RABBITBRUSH—RUBBER RABBITBRUSH
Chrysothamnus nauseosus ssp. hololeucus
Flower: Yellow

Sunflower Family - *Asteraceae*

This bush, in bloom, is a fall spectacular. It lines the roadsides and washes in the East Mojave, sometimes in dense stands, as in Cedar Canyon Wash near Government Holes, and provides a stunning visual display for a month or so, beginning in late September.

The shrub grows to be about 5 feet tall and has almost white, flexible stems. The new tiny green leaves which sprout in early spring are very conspicuous against the light-colored stems. They turn gray as they mature. Leaves are linear, about 2 inches long, with turned-up margins.

Golden Rabbitbrush is a member of the largest plant family in North America, well represented in the Scenic Area.

The small yellow flowers grow in dense clusters at the ends of stems. Each bloom by itself is insignificant and has only disk flowers, no rays, with outer bracts in 5 vertical rows.

It is sometimes called Rubberbrush because of its high rubber content.

Generally I think the Latin species name "nauseosus" is a misnomer because the scent is rather pleasing. But after a rain I walked through a "forest" of these plants and found the smell to be, indeed, nauseating.

You'll often see pea-size silvery gall growths on these plants. If you cut one open, you'll find a small grub inside.

LINEAR-LEAVED GOLDENBUSH
Ericameria (Haplopappus) linearifolia
Flower: Yellow

Sunflower Family - *Asteraceae*

This is a woody perennial, much-branched shrub, usually 2 or 3 feet high. Leaves are alternate, often in bunches, 1 inch long and very narrow, gland dotted and resinous. They turn a gray color in winter. The plant "greens up" in early March.

Bloom begins in late March. The sunflower-type blooms have both disk and ray flowers. Rays are bright yellow, about 1/2 inch long.

This species is common in the higher elevations up to 6,000 feet. Look for it on the hillsides of Cedar Canyon near Government Holes and at Camp Rock Springs.

Linear-Leaved Goldenbush and the following 3 shrubs were formerly in the Haplopappus genus.

A small, rounded bush with similar leaves creates a golden haze over much of the East Mojave. **COOPER GOLDENBUSH** (*Ericameria cooperi*) is common in Lanfair Valley. The flower heads are mainly yellow disk flowers with 1 or 2 tiny, randomly-placed ray flowers. It looks as if a strong wind might have blown off the other rays. Bloom is from March to June.

TURPENTINE BRUSH (*Ericameria laricifolia*) is a rounded shrub, 2 or 3 feet tall, with small green linear leaves covered with tiny shiny dots (use a hand lens). It is a fall bloomer found on rocky slopes between 3,500 and 6,500 feet elevation. Small yellow heads have both ray and disk flowers and densely cover the bush.

ROCK GOLDENBUSH—WEDGELEAF GOLDENBUSH (*Ericameria cuneata*) is a little woody, evergreen shrub found in the granite outcroppings of the higher elevations. Our own Quail Rock has Rock Goldenbush high on its flanks and growing at the base of the rock.

The crowded, 1/2-inch-long, wedge-shaped leaves remain on the plant all year, turning slightly gray in winter. They are deep green, resinous, gland-dotted and delightfully scented. The flower heads grow in small clusters, made up of mainly yellow disk flowers. Bloom occurs in fall.

Linear-Leaved Goldenbush

Cooper Goldenbush

Rock Goldenbush is a fall bloomer

MATCHWEED—SNAKEWEED—ROSIN WEED
Gutierrezia microcephala
Flower: Yellow

Sunflower Family - *Asteraceae*

No plant fits the description of "rounded" as well as Matchweed. This little perennial subshrub covers many acres of the East Mojave yet mainly goes unnoticed because of its insignificant size. It's an impressive sight, however, when whole hillsides of Matchweed come into bloom in fall.

New leaves start appearing in March on the dried-out stems still sporting straw-colored seed heads from the last bloom.

The compact bush is usually about 1 foot high with many stems. Bright lime-green, threadlike leaves are about 1/2 inch long, arranged alternately. Flower heads are tiny but very numerous. The heads have 1 or 2 ray flowers and 1 or 2 disk flowers.

A handful of the resinous branches will start a campfire with just one match.

Spanish Americans make a tea from the leaves to drink and also bathe in it to relieve aches and pains.

This plant is rarely eaten by livestock and will "take over" on heavily grazed land.

As you drive along Cedar Canyon Road, east past Black Canyon Road, you'll see Matchweed covering the alluvial slopes of Pinto Mountain to the north.

MOJAVE HORSEBRUSH—FELT THORN
Tetradymia stenolepis
Flower: Yellow

Sunflower Family - *Asteraceae*

Mojave Horsebrush is a rigid, gray, spiny shrub, locally called Gray Sage, growing to about 3 feet high. The strawcolored 1 inch spines are placed alternately on the stems at fairly regularly spaced intervals. The linear leaves, 1/2 inch long, are covered with dense white hairs and are a soft gray. The margins curl upward to form a deep trough. They grow in alternate bundles. New spring growth produces soft, gray, flexible spines which harden with age.

Yellow blooms consist of disk flowers only, growing at the tops of stems, 5 per head, with woolly white bracts. Flowers occur in summer.

There are good specimens growing along the fence line on Cedar Canyon Road about 2 miles east of Kelso-Cima Road, near Hole-in-the-Wall, and in the Granite Mountains.

MORMON TEA—BRIGHAM TEA—JOINT FIR
Ephedra spp.
Flower: Yellow

Joint Fir Family - *Ephedraceae*

Ephedras belong to that division of plants—**CONIFEROPHYTA**—that bear cones.

There are several species of Ephedra in the Scenic Area. These are green shrubs made up of many jointed stems, usually leafless. The male and female cones are on separate plants and develop at the joints on the stems in spring. *Ephedra viridis* (Latin, meaning "green") is bright green with thin, erect, broomlike branches. Good specimens can be found at Rock Springs. *Ephedra nevadensis* has bluish-green branches. Both have leaf nodes arranged in twos on the stems. Other species have the nodes in threes.

Mormon Tea has been used historically throughout the West. As the common name implies, early Mormon settlers brewed the branches into a refreshing beverage. The tea has diuretic properties and is used by Native Americans in the Southwest for stomach ailments. The branches can be picked at any time of year, cut into small pieces, and steeped in boiling water for 15 minutes. We enjoy a cup of this tea quite often. Cattle eat ephedra, but only after a freeze in winter.

PAPERFLOWER
Psilostrophe cooperi
Flower: Yellow

Sunflower Family - *Asteraceae*

Paperflower is an unusual little bush because it does not lose its flowers. They age to a papery straw color and remain on the plant for months.

This perennial subshrub forms a compact, rounded silhouette about 12 to 18 inches high. It has many stems and gray-green linear leaves. The flowers are solitary at the ends of the upper branches. There are 4 to 8 bright yellow, 1/2-inch-long ray flowers, each 3-lobed on the outer margins, and 4 to 20 disk flowers.

Paperflower begins blooming in April and, depending on rainfall, again in fall.

It is common above 2,000 feet. Large stands grow along Cedar Canyon Road near Kelso-Cima Road and on Essex Road near Black Canyon.

PIÑON WORMWOOD
Artemisia ludoviciana
Flower: Yellow

Sunflower Family - *Asteraceae*

This delicate perennial is a close cousin to Big Sagebrush. Like its larger relative, it is very aromatic when crushed and also tastes bitter. The leaves and stems are covered with white hairs.

Tiny flowers bloom on slender stems. Most of the plant dies back in winter. New growth starts early in spring from rhizomes. The plants I've encountered are from 1 to 2 feet tall.

The common name of wormwood for plants in this genus comes from the belief they will help get rid of pinworm and roundworm infections.

Piñon Wormwood grows among rocks in the higher elevations, up to 7,000 feet. There are many nice plants tucked in among the big boulders lining the wash behind Rock Springs.

PRINCE'S PLUME
Stanleya pinnata ssp. pinnata
Flower: Yellow

Mustard Family - *Brassicaceae*

You can't miss this showy plant if you include a visit to Ft. Piute on the old Mojave Road during your stay in the Scenic Area. Piute Creek is bedecked, as if for a grand party, with these 4 or 5-foot-tall perennials sporting waving yellow-tipped wands. You'll also find it near Granite Pass on Kelbaker Road. Great numbers of this plant grow on top of Table Mountain, especially in the burned-off section.

There are several flexible stems with gray-green, deeply-cleft leaves up to 8 inches long, growing mainly on the lower portions. The yellow flower clusters are at the tops of stems. Each small flower in the cluster has 4 petals about 1/2 inch long and 4 yellow sepals. Fruit is a slender pod up to 3 inches long. These plants prefer soil with heavy concentrations of selenium and are probably poisonous.

Prince's Plume grows at elevations between 1,000 and 6,000 feet. Bloom is from May to July.

RAYLESS ENCELIA—BUSH ENCELIA
Encelia frutescens
Flower: Yellow

Sunflower Family - *Asteraceae*

Lining Kelbaker Road, this rounded, much-branched shrub comes into bloom as early as March in the lower elevations. Dark green, 1 inch long, 1/2 inch wide leaves grow alternately on white stems. The plant is usually 3 or 4 feet high.

Unlike other encelias, blooms consist of disk flowers only, about an inch across. They have a pleasant fragrance.

If there is adequate rainfall these shrubs will bloom again in fall.

SANDWASH GROUNDSEL—THREADLEAF GROUNDSEL
Senecio douglasii var. monoensis
Flower: Yellow

Sunflower Family - *Asteraceae*

Bright green threadlike leaves and sunny daisy-like yellow flowers enliven the desert washes where this bushy perennial prefers to grow.

Flower heads are about 1 inch wide with both ray and disk flowers. There are 10 to 13 yellow rays.

The blooming season is from March to May with sometimes a second bloom in the fall, after which the leaves drop off.

Sandwash Groundsel is toxic to cattle and is avoided, thus it increases on grazed land.

Look for this plant along Cedar Canyon Road near Kelso-Cima Road, on Kelbaker Road north of I-40 near the Granite Mountains and on Black Canyon Road near Hole-in-the-Wall.

SQUAWBUSH—SQUAWBERRY
Rhus trilobata var. anisophylla
Flower: Yellow

Sumac Family - *Anacardiaceae*

Squawbush is a relative of Poison Oak, but those of you who fear those dreaded 3 leaflets can rest easy around this shrub as it is not poisonous. Leaves on Squawbush are made up of 3 leaflets—2 lateral leaflets and a terminal leaflet which is the largest and is 3-lobed. The terminal leaflet does not have a stalk which Poison Oak always has.

Most parts of this plant are, or have been, used for food, fuel, paint dyes, basketry and medicines. Native American women used the stems for plaiting baskets, hence the common name of Squawbush.

Flowers appear in March and April, before the leaves. Blooms are in small clusters. The 5 yellow petals are less than 1/8 inch long. Red berries ripen in summer and are favorites of local birds and mammals. The fruits are "drupes"—fleshy, one-seeded stone fruits like plums. They are quite tasty, if a bit tart. I make a lemonade-type drink from them. A local name is California Christmas Berry.

Squawbush is deciduous but you can learn to differentiate it from other bare leafless shrubs by the slightly reddish cast of the bark. The stem ends usually point downward a little. It grows to about 5 feet tall.

Look for Squawbush near granite outcrops on Black Canyon Road about 5 miles south of Cedar Canyon Road. It's common above and below Rock Springs.

VIRGIN RIVER ENCELIA
Encelia virginensis
Flower: Yellow

Sunflower Family - *Asteraceae*

   Blooms, with yellow ray and disk flowers, are atop long naked stems on this much-branched shrub. Green, somewhat oval, leaves are 1/2 inch to 1 inch long and 1/2 inch wide. There are 12 to 20 ray flowers. This Encelia is widespread in the Scenic Area, up to 5,000 feet elevation, and prefers gravelly washes and hillsides. Many grow near Ft. Piute. Bloom is in April and May.

WOOLLY-FRUITED BURSAGE—BURBUSH
Ambrosia eriocentra
Flower: Yellow

Sunflower Family - *Asteraceae*

Here is a close relative of Burrobush but a much larger shrub, up to 4 feet tall, with slender white stems. Green leaves are about 1 inch long, toothed or lobed. They're sticky when crushed. The plants are fairly late to leaf out. Inconspicuous male and female flowers grow on the same plant. The female heads form a small 1/4 inch bur which is covered with long, cottony, white hairs.

The leaves are very aromatic. I've used them for seasoning and hang a bundle of them in the bathroom.

Woolly-Fruited Bursage lines both sides of Cedar Canyon Road east of Kelso-Cima Road for a short distance. Good specimens grow in the Granite Mountains and in the wash below Rock Springs.

Leaves

Cottony burs

ARROWWEED
Pluchea sericea
Flower: Light purple

Sunflower Family - *Asteraceae*

The slender silvery stems of this tall, willowlike shrub were used for arrow shafts by Native Americans. Silky gray, narrow leaves, about 1 inch long, are crowded on stems up to 12 feet high. Light purple disk flowers grow in clustered heads on the ends of stems. Bloom is from March to July.

Arrowweed likes moist places and is common in the low desert along the Colorado River near Needles. In the Scenic Area you'll find it blooming in May along Piute Creek.

BLUE SAGE—PURPLE SAGE—GRAY BALL SAGE
Salvia dorrii ssp. dorrii
Flower: Blue

Mint Family - *Lamiaceae*

Here is a small shrub as pretty in winter as when in bloom. It is truly an "evergray." The silvery leaves are deliciously aromatic. I use them in cooking often and enjoy chewing on a leaf occasionally. It was of this sage Zane Grey wrote in "Riders of the Purple Sage."

Blue Sage (or Purple Sage) is a broad shrub, growing to about 3 feet tall. Leaves are opposite, 1/2 inch to 1 inch long, broadest near the tip, sometimes growing in clusters. They are a beautiful blue-gray color contrasting with the more aqua-gray of Big Sagebrush, a common associate. Leaf margins curl upward to form a trough, then tip downward.

Small blue flowers occur in whorls. Each flower is 2-lipped with 2 bright yellow-tipped stamens and 1 pistil arching outward. The whorls are usually 2 to 4 "stories" high on the stems, like the whorls of Chia but spaced more closely together. Purplish-green bracts are underneath the flower clusters. Before the flowers open the bracts give the plant a purple cast that changes to blue when full bloom occurs.

Blue Sage grows above 2,500 feet elevation. It lines the roadsides of Black Canyon heading south from Cedar Canyon. There are many on the Teutonia Peak Trail.

INDIGO BUSH
Psorothamnus (Dalea) fremontii var. fremontii
Flower: Purple

Pea Family - *Fabaceae*

Deep royal purple, pea-type flowers are a striking contrast to the light gray-green stems and leaves on this handsome 3 to 5-foot-tall shrub. Several species of Indigo Bush, close relatives of Smoke Tree, grow in the lower Colorado Desert and we are fortunate to have this one at higher elevations in the Scenic Area.

You can find Indigo Bush near Kelbaker Road just north of Interstate 40 and near the intersection of Black Canyon Road and Essex Road. Bloom is in April and May.

MOJAVE SAGE
Salvia mojavensis
Flower: Light blue

Mint Family - *Lamiaceae*

This sage grows only in the Mojave Desert (an endemic) and is well-represented in the Scenic Area.

Pale blue flowers contrast starkly with the deep green leaves. This, and the other sages, are popular bee plants.

Mojave Sage grows to about 2 feet high and is usually a little broader. Opposite leaves are about 3/4 inch long, wrinkled above with a network of veins below and strongly aromatic.

Flowers, growing in whorls, are deeply lobed topping a pale slender tube, about 1 inch long, with outer white membranous bracts. Stamens stick out about 1/2 inch above the corolla. A single cluster blooms at the end of a branch. Flowering occurs in late spring-early summer.

Mojave Sage likes rocky canyons and washes. It can be found in the washes near the Mitchell Caverns-Bonanza King Mine areas of the Providence Mountains. There are several in the side wash on the Cedar Canyon Wash Walk. Hackberry Mountain supports large populations of these handsome shrubs.

PAPERBAG BUSH—BLADDER SAGE
Salazaria mexicana
Flower: Purple

Mint Family - *Lamiaceae*

The hundreds of little inflated "paper bags" on this shrub attract attention. They detach at maturity and are blown across the desert to disperse their seeds.

Paperbag Bush, sometimes locally called Bottle Brush, is usually about 3 feet high with intricate, interlaced branches, often with flexible spines. Small 1/2 inch leaves are placed opposite on the stems. The purple flowers are 2-lipped; the upper lip deep purple, the lower cream colored. Each flower is about 3/4 inch long.

The calyx swells and becomes a bladder-like seed case, about 3/4 inch wide, containing 4 nutlets. A purple aura seems to surround the plant during bloom and fruit development. Antelope ground squirrels climb into the bushes to reach the bags.

The bush is deciduous, dropping its leaves fairly early, but the old bags may stay on the plant until the following spring bloom which occurs from April to June.

Paperbag Bush is common along roadsides and washes. It lines Cedar Canyon Road a few miles east of Kelso-Cima Road and Kelbaker Road near the Granite Mountains.

RANGE RATANY—LITTLE-LEAVED RATANY
Krameria parvifolia imparata
Flower: Purple-red

Krameria Family - *Krameriaceae*

This twiggy, much-branched little bush lies close to the ground, usually not more than a foot high, spreading outward 3 or 4 feet. Much of the year it appears quite lifeless.

The leaves are narrow, gray and up to 1/2 inch long. Flowers are at the ends of 1/2-inch-long peduncles. There are 5 sepals which are purple on the inside surface. Flowers have 3 upper petals smaller than the sepals and 2 reduced lower petals. Fruit is a round, spiny, barbed pod.

Ratany sparkles when the sun glistens on the shiny reddish-purple flowers.

It is common throughout the East Mojave. On the Cedar Canyon Wash Walk you'll see many of these shrubs on the hillsides.

There is another Ratany in the Scenic Area—**WHITE RATANY** (*Krameria grayi*)—which is similar to Range Ratany, except the fruit has spines which are barbed only at the tip, umbrella fashion. You may see it on the Mojave Road-Ft. Piute Hike.

Flowers and spiny fruit of Range Ratany

SPINY HOPSAGE
Grayia spinosa
Fruit: Pink, Red, Purple, Green

Pigweed Family - *Chenopodiaceae*

Spiny Hopsage is a close cousin to the saltbushes (Four-Wing Saltbush, Desert Holly). The main distinction is in the fruiting bracts which on Spiny Hopsage are completely joined to form open sacs around the fruit. These fruit sacs are about 1/2 inch long and vary in color from green to slightly pink to deep purple.

This is a stiffly-branched shrub about 2 to 4 feet tall with a few spines. The gray-green leaves are 1/2 inch to 1 inch long. Male and female flowers grow on separate plants and are very inconspicuous. The female plant is attractive when covered with clusters of delicate fruiting bracts surrounding a tiny seed. Bloom is from April through June. This is a valuable forage plant. Local ranchers call it Apple Brush.

You can find Spiny Hopsage along Cedar Canyon Road, along Kelbaker Road near Interstate 40 and near Cima Road.

Spiny Hopsage - female plant

TURPENTINE BROOM
Thamnosma montana
Flower: Purple

Rue Family - *Rutaceae*

Turpentine Broom earns its name because of its broom-like branches, leafless most of the year, and the very strong scent of its stems and fruits. Local ranchers call this plant Skunk Weed. The Rue Family includes citrus.

It is yellow-green and from a distance may be mistaken for Mormon Tea. This shrub grows up to 2 feet tall and about as wide. Tiny linear leaves fall off after the shrub flowers.

The blooms are quite unique—a deep purple, almost black. The flowers have 4 petals but are arranged in a tube, no more than 1/2 inch long, that does not open up wide. There are 4 long stamens and 4 short. The style barely sticks out of the flower tube. Flowering occurs in spring. Fruit and flowers may be on the bush at the same time. Fruit is a 2-lobed, pea-size capsule covered with gland dots, with a very strong odor.

Turpentine Broom is common in the Scenic Area. Look for it along Black Canyon Road near Hole-in-the-Wall, along Cedar Canyon Road 3 or 4 miles east of Kelso-Cima Road, along Wildhorse Canyon Road, and at Mid-Hills Campground.

Flowering and fruiting stem of Turpentine Broom

Paperbag Bush

Desert Five Spot

Turpentine Brush in front of Blue Yucca

Yellow Tack-Stem

Indigo Bush

Lilac Sunbonnet

Silver Cholla

Bigelow Mimulus

Indian Paintbrush

Spiny Hopsage

Black-Banded Rabbitbrush

Giant Four O'Clock

Canterbury Bell

Mariposa Lily

Hedgehog Cactus

BROWN-EYED PRIMROSE
Camissonia clavaeformis
Flower: White

Evening Primrose Family - *Onagraceae*

You'll find this small-flowered member of the Evening Primrose Family growing in the same habitats, below 4,000 feet elevation, as Yellow Cups. One or several reddish stems with green toothed leaves grow up to 1 foot tall. Flowers are clustered at the tops of stems. The blooms consist of 4 white petals, each with a brown spot at the base. Bloom is in early spring. Club-shaped seed capsules are about an inch long.

COYOTE TOBACCO
Nicotiana attenuata
Flower: White

Nightshade Family - *Solanaceae*

Cedar Canyon Wash, between Government Holes and Rock Springs, has many Coyote Tobacco plants. They are usually 2 to 4 feet tall.

Leaves at the base of the plant are large—up to 4 or 5 inches—with long, 1 or 2 inch petioles (leaf stalks), but grow progressively smaller toward the top of the plant. They're smelly when crushed.

The white, 1-inch-long flowers are tubular shaped. Bloom is from May to October.

The plant contains nicotine. The Latin name comes from Jean Nicot, a Frenchman who introduced tobacco to France. Indians dried and smoked the leaves. They were also used as poultices.

Another tobacco plant grows in the Scenic Area, below 4,000 feet elevation. **DESERT TOBACCO** (*Nicotiana trigonophylla*) has a similar white, 5-lobed, tubular flower but its leaves have no petioles. They clasp the stems. This plant grows up to 2 1/2 feet tall. Bloom is from March to June. Look for Desert Tobacco on the Mojave Road-Ft. Piute Hike.

Coyote Tobacco

DESERT CHICORY
Rafinesquia neomexicana
Flower: White

Sunflower Family - *Asteraceae*

Along Kelbaker Road in April many of the small plain Burro-bush plants are decorated with white showy flowers. These are blooms of Desert Chicory poking out of the protective, supportive branches of the little shrub.

The stems of this pretty annual are weak and it grows best with support from other plants. Leaves are mainly at the base, 2 to 8 inches long, and deeply lobed. Pure white 1-1/2-inch-wide inflorescences consist of ray flowers only (like dandelions), often with light purple veins underneath.

The common name comes from the fact that its bloom resembles the more common, blue Chicory used for beverages.

DESERT LILY—AJO LILY
Hesperocallis undulata
Flower: White

Lily Family - *Liliaceae*

It's quite unexpected to find lovely white lilies, much like our Easter Lilies, growing in desert sands. This attractive plant grows only in low, sandy deserts, below 2,500 feet elevation, and can be seen blooming from March to May along Kelbaker Road east of Baker and near Kelso Dunes.

Spring growth is from a deeply buried bulb which Native Americans and Spaniards used for food. The flavor must resemble garlic which is "ajo" in Spanish. Long, blue-green leaves with wavy margins (undulata is Latin for "wavy") grow just above the ground. The flower stalk may be 2 or more feet high. The 2-inch-long flowers have a green stripe on the back of each petal.

DESERT MILKWEED
Asclepias erosa
Flower: White

Milkweed Family - *Asclepiadaceae*

This perennial herb has thick white juice, opposite leaves, small white flowers growing in big rounded clusters, and seeds, tufted with silky white hairs, in large pods. The leaves are wide and long, covered with short woolly hairs. Their veins are arranged in a fishbone pattern. Each flower of Desert Milkweed is made up of 5 petals joined together (the corolla) and has 5 sepals. The blooming period begins in May. Plants are usually 2 to 4 feet tall. In the fall of 1990 I found a 6-foot specimen in Cedar Canyon Wash. The dried stalk was still standing the following spring.

Flowers attract many tarantula hawks (large black and orange wasps). Leaves are eaten by big, fat, yellow-black-white-banded caterpillars. These are the larvae of Monarch butterflies.

Desert Milkweed is common on the roadsides of Kelbaker Road, Lanfair Road, and Cedar Canyon Road.

Pods of Desert Milkweed

DESERT PRIMROSE—TUFTED EVENING PRIMROSE
Oenothera caespitosa
Flower: White

Evening Primrose Family - *Onagraceae*

This delicate looking Evening Primrose (not related to our garden primroses) has 4 white petals, each slightly heartshaped. The flowers open in the evening (hence the family name) and age to light pink the following day.

The young plants grow from perennial roots and form a thick basal rosette of deep green leaves, sometimes tinged purple. Leaves are 3 or 4 inches long, broader at the base, tapering to a point at the top. They are hairy and have deep wavy margins.

Flowers are sweet smelling. They are solitary, on short stems, each with 4 petals, 4 sepals and 8 stamens. As with many white, night-blooming flowers, they are pollinated by moths.

The seed capsules on Desert Primrose are about 1 1/2 inches long, fat and warty, and form in the center of the basal leaves. The capsule splits into 4 parts to empty its contents. These plants like to tuck up next to rocks or grow on the banks of washes. Bloom is in spring and summer.

The low-spreading plant with deeply lobed gray-green leaves and similar large white flowers common in Cedar Canyon is **WHITE EVENING PRIMROSE** *Oenothera avita*. Its seed capsules are narrow, about 1 1/2 inches long. The blooms are not scented.

An Evening Primrose you may find on Kelso Dunes is **DUNE PRIMROSE** or **BIRDCAGE PRIMROSE** (*Oenothera deltoides*). It blooms from March to May. When the plant dies the stems curl up and inward to form a dried basket or "birdcage" which will last for many years. The larvae of a sphinx moth eats this plant.

There are many species of Evening Primrose in the desert. Kelso-Cima Road truly becomes a "Primrose Lane" in spring.

Desert Primrose

Sprawling mats of White Evening Primrose

DODDER
Cuscuta spp.
Flower: White Stems: Orange

Dodder Family - *Cuscutaceae*

Masses of shiny orange, string-like stems are the characteristic feature of this strange parasitic plant. It has tiny flowers and produces seeds. After the seedlings find a host plant and attach their stems to the tissue of that plant, Dodder's roots shrivel and it becomes entirely dependent on the host and may eventually kill it.

Dodder is not too common in the Scenic Area but when seen will arouse curiosity. It prefers sandy wash areas, usually at the lower elevations, although we occasionally have it entwined around plants at Quail Rock, at 5,000 feet elevation. You may find it near Kelbaker Road heading east from Baker or near Interstate 40. South of I-40, just outside the Scenic Area, I've seen acres of bushes entangled with this orange, spider web-like parasite.

Dodder entwined on Desert Trumpet

FREMONT PINCUSHION
Chaenactis fremontii
Flower: White

Sunflower Family - *Asteraceae*

This annual may grow 15 inches tall, with many branches and many blooms, in a year with adequate rain. It is often much smaller.

The flower head consists of tiny, 5-lobed white blooms with those at the outer edges being larger.

Fremont Pincushion is early to flower, often blooming by mid-March. Look for it along Black Canyon Road near Essex Road and near Kelso.

LILAC SUNBONNET—SPOTTED GILIA
Loeseliastrum setosissima ssp. punctata (Langloisia punctata)
Flower: White

Phlox Family - *Polemoniaceae*

Five white, purple-dotted corolla lobes and narrow bristle-toothed leaves distinguish this low-growing annual. Punctata is Latin for "dotted." The flowers are about 1 inch wide. I found this plant blooming in April near a wash on the Volcanic Rock Hill Walk.

NEW MEXICO THISTLE—DESERT THISTLE
Cirsium neomexicanum
Flower: White

Sunflower Family - *Asteraceae*

Here is another member of the large Sunflower Family. It's easily recognized from the prickles on the leaves and stems. It is a biennial plant. The first year it forms a basal rosette of leaves. The second year it sends up stems and flowers.

This species has white or pink-tinged blooms, in heads up to 2 inches wide, usually solitary, made up of disk flowers only. They bloom in late spring.

The plants are 2 to 5 feet tall and grow in rocky soil at elevations of about 3,000 to 6,000 feet. You should be able to find New Mexico Thistle in the Rock Springs area.

PALMER PENSTEMON—SCENTED PENSTEMON—BALLOON FLOWER—BEARD-TONGUE—SNAPDRAGON
Penstemon palmeri
Flower: White, Pink

Figwort Family - *Scrophulariaceae*

The sterile stamen in this flower (see Eaton Firecracker) protrudes out of the corolla and is heavily bearded, hence the common name of Beard-Tongue.

Palmer Penstemon is a tall, erect perennial plant, sometimes growing to 5 or 6 feet. The tops of the stems bend down while growing but straighten up when flowering occurs. Light green, opposite leaves have sawtooth edges, and grow only on the lower half of the stems. The white balloon flowers have two lips. The corolla is tinged with pink and has deep pink "guide lines," leading to the throat, on the lower lip. These lines direct insects to the pollen. Another attraction for insects (and people, too) is the sweet smell of the flowers. The flowering inflorescence may be 2 feet high, above the leaves.

This penstemon is found near washes in the Providence, New York, Granite and Clark Mountains. There is a reliable group along Cedar Canyon Road 2 miles east of Black Canyon, blooming in May and June.

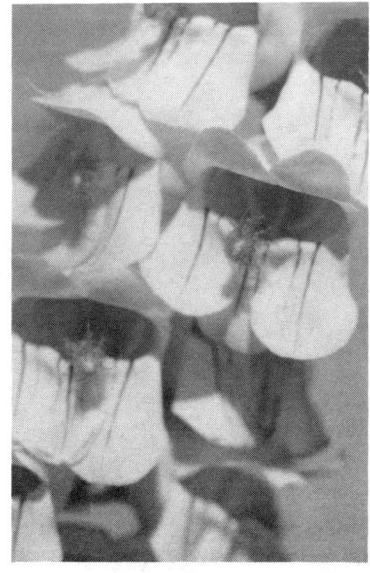

PRICKLY POPPY—COWBOY'S FRIED EGG
Argemone corymbosa
Flower: White

Poppy Family - *Papaveraceae*

White, wrinkled, tissue paper flowers with bright yellow centers, and a red spot in the exact middle, on plants up to 3 feet high, will easily catch your eye. Stems and leaves of this perennial herb are covered with prickles. The stems ooze an orange sap when broken.

Prickly Poppy is common on the roadsides near Kelso. Bloom is in April and May.

PUNCTURED BRACT—SAUCER PLANT
Oxytheca perfoliata
Flower: White

Buckwheat Family - *Polygonaceae*

Here's a small member of the Buckwheat Family with such unusual characteristics it is very easy to identify and a real treat to find. The slender stems on this little annual pass through cup-shaped bracts. Basal leaves are about 1 1/2 inches long. The short main stem branches 2 or 3 times horizontally. Flowers are tiny, at the ends of branches. Bloom is from April to July. I found Punctured Bract along Cedar Canyon Road near Kelso-Cima Road. It prefers sandy, gravelly soils up to 6,000 feet elevation.

RATTLESNAKE WEED—WHITE MARGINED SPURGE
Chamaesyce (Euphorbia) albomarginata
Flower: White

Spurge Family - *Euphorbiaceae*

Only a rattlesnake slithering by on its belly can get a good view of this tiny plant, but that's not how it came by its common name. It was once used to treat snake bites.

The ground-hugging mats often grow in such profusion you will surely notice them while walking through the desert. The small green leaves are placed opposite on the sides of 2 to 10-inch-long stems which have white sap. What appears to be the flowers are actually white appendages resembling petals with a dark spot on the base. Each "flower" is a small white circle with a dark purple center. It requires a hand lens to see the actual flower set inside. Rattlesnake Weed blooms from May to November and then drys up.

It is common in the Scenic Area. Look for it at Rock Springs or lining Kelbaker Road near the Granite Mountains.

SACRED DATURA—JIMSON WEED—THORN APPLE
Datura wrightii (meteloides)
Flower: White

Nightshade Family - *Solanaceae*

Most desert plants have developed unique adaptations for severe climatic conditions such as spines, leaf drop, short lives, etc. Now meet the show-off of the arid West.

Big 4 or 5-inch-long green leaves spill over the roadsides in large clumps; massive white funnel-shaped flowers give the appearance of an unending water supply. Because of the large flowers and leaves, and because it is a quite common roadside plant, Datura causes more curiosity among visitors than most plants.

Sacred Datura has perennial roots. After the flower and fruit cycle is complete, the plant dies but usually remains as a skeleton, with dangling spiny seed capsules, until the next season when new growth pushes its way up through the dead stems.

The funnel-shaped flowers are white, often tinged with purple, and uncurl from a tubular calyx. Sometimes the flowers close during the afternoon.

All parts of the plant contain poisonous alkaloids. Native Americans used it to produce hallucinogenic visions, but in untrained hands it can be lethal and should never be taken internally.

The common name "Jimson Weed" is a shortened form of "Jamestown Weed" where it poisoned soldiers in colonial days. The name "Thorn Apple" describes the seed capsule—a prickly, round, golf ball-size container drying from light green to brown.

Other members of the Nightshade Family include tomato, potato, tobacco, petunia and bell pepper.

SPECTACLE POD
Dithyrea californica
Flower: White

Mustard Family - *Brassicaceae*

Tiny pairs of eyeglasses grow on this annual of the low sandy deserts. The plant has several stems, usually under 1 foot high, and alternate leaves, 1 or 2 inches long, with wavy, toothed margins. Small, white, 4-petaled flowers grow in dense clusters at the tops of stems. The fruit is an oval green pod, less than 1/2 inch long, notched in the center above and below, which resembles a pair of spectacles or, from the Greek word dithyrea, "two shields."

Spectacle Pod can be found on the sandy roadsides near Kelso blooming from March to May.

TECOPA SKELETON WEED
Eriogonum deflexum brachypodum
Flower: White, Pink

Buckwheat Family - *Polygonaceae*

Nature provides us with a handsome roadside dried-plant arrangement when these wide-spreading, flat-topped annuals turn reddish-brown in fall. Most members of the Buckwheat Family turn a russet color when dead.

Skeleton Weed, so named because it may remain dried in place for over a year, is usually about 10 inches tall with round basal leaves. The stems branch to form flat crowns a foot or so across. Tiny, rice-sized flowers hang from the spreading horizontal branches. Bloom is from May through summer.

Skeleton Weed is common on roadsides in the Scenic Area. A good display is usually found on Cedar Canyon Road just east of Kelso-Cima Road and on Black Canyon Road near Hole-in-the-Wall.

TIDY FLEABANE
Erigeron pumilus ssp. concinnoides
Flower: White, Pink, Lavender

Sunflower Family - *Asteraceae*

Gravelly slopes, at elevations between 3,500 and 6,000 feet, are home to this perennial herb. It is usually 6 to 10 inches tall with several hairy stems and linear hairy leaves. Delicate blooms have many (50 to 100) thin ray flowers surrounding yellow disk flowers. Buds nod gracefully before opening. Bloom is in May and June.

It was once used to get rid of fleas, hence the common name.

Rock Springs-Government Holes area and Mid-Hills are good spots to look for Tidy Fleabane.

TUMBLEWEED—RUSSIAN THISTLE
Salsola iberica (australis)
Flower: White

Pigweed Family - *Chenopodiaceae*

Meet the one plant growing in the East Mojave (and most everywhere else) I could do without. It is, indeed, a pigweed. This non-native annual plant scatters its seeds across the desert with every wind.

In spring the new growth is tender and edible. Leaves and flowers are small. The blooms are greenish-white tinged with red. Before the plant dries and becomes all-over prickly, cattle and mule deer relish the tender growth. Unfortunately, they don't get it all and, as the plant matures, the scalelike leaves become very sharp so it is no longer palatable. It dries to a rounded clump that is pulled loose by the wind.

I include Tumbleweed for the simple reason that it, regrettably, is here. Look for it along roadsides near Kelso and in disturbed sites almost anywhere.

WHITE ASTER
Leucelene ericoides
Flower: White

Sunflower Family - *Asteraceae*

Delicate white ray flowers surround a bright yellow center on this little plant that grows just 4 to 6 inches high. It's a perennial from a woody base, with many stems covered with tiny, hairy leaves. There are many ray flowers, each about 1/4 inch long, forming the daisy-like bloom. The flowering period is from April through summer.

White Aster prefers rocky, dry slopes in Piñon-Juniper country of the Providence, New York and Clark Mountains. You're sure to find it blooming in May in Caruthers Canyon or near Rock Springs.

WHITE TIDY TIPS
Layia glandulosa
Flower: White

Sunflower Family - *Asteraceae*

The neatly 3-lobed white petals on this short spring-flowering annual are crisply "tidy." The plant is usually 5 to 10 inches tall, with one or several branches, and in the East Mojave does not seem to grow in dense colonies but is scattered throughout an area. The stem is tinged reddish-purple. Basal leaves are notched or toothed but the few slim tapering upper leaves have even margins. Both stem and leaves are hairy.

Tidy Tips has both ray and disk flowers. The head is about an inch wide. Bloom is from March to June. Look for Tidy Tips on the Teutonia Peak Trail Walk and near Rock Springs. It's common throughout the Scenic Area.

WISHBONE BUSH
Mirabilis bigelovii
Flower: White

Four O'Clock Family - *Nyctaginaceae*

This smaller relative of Giant Four O'Clock has a branching pattern like a wishbone. Leaves are opposite on the branches. They are about an inch long, green, covered with soft hairs and slightly sticky. The white calyx tube is 5-lobed, set in a cuplike involucre. The flowers close up by mid-day. Bloom is in spring. It's a large-rooted perennial that dies back in winter.

Wishbone Bush grows in rocky canyons and on hillsides. Look for it near Rock Springs and Ft. Piute.

CHINCH-WEED
Pectis papposa
Flower: Yellow

Sunflower Family - *Asteraceae*

Summer rains stimulate the growth of this little strongly-scented annual. Thin, yellow-green stems grow about 8 inches tall with narrow, linear leaves. The gland-dotted involucres are topped by 7 to 9 yellow rays and 10 to 15 disk flowers. The area around Camp Rock Springs is often carpeted with Chinch-Weed after the usual August thunderstorms of the East Mojave.

COYOTE MELON—COYOTE GOURD
Cucurbita palmata
Flower: Yellow

Gourd Family - *Cucurbitaceae*

"Only good enough for coyotes," was the opinion of the Native Americans for the food value of these gourds.

The leaves, flowers and fruits of this sprawling plant are quite conspicuous along disturbed roadsides where they are common throughout the East Mojave below 4,000 feet elevation.

Spring growth from deep tubers is very rapid. Stems spread out from 3 to 6 feet. Leaves are palmately 5-lobed. Male and female flowers grow on the same plant, the female being larger. Yellow blooms are trumpet-shaped with 5 lobes. Fruit is a round green gourd, drying to light yellow, up to 4 inches in diameter. When the stems and leaves dry up, the gourds remain attached by a woody stem and are very noticeable.

I enjoy making rattles from the old gourds—inserting a stick through holes drilled at the top and bottom and painting them with ancient petrogylph designs. The seeds rattle inside when I shake the stick.

DESERT DANDELION
Malacothrix glabrata
Flower: Yellow

Sunflower Family - *Asteraceae*

Desert Dandelions are cheery annuals much like our common lawn dandelions except these usually have red centers. They form a golden carpet over large areas in years of good rainfall, each plant sending up 10 or more flowers.

Malacothrix means "soft hair" in Greek and refers to the seed heads. Glabrata is "bald" in Latin and refers to the hairless leaves.

Leaves are up to 5 inches long in threadlike divisions.

Flowers may be 2 inches across. They consist of rays only—no disks. Bloom is from March to June throughout Creosote Bush country. Kelbaker Road out of Baker is lined with Desert Dandelions in spring.

SCALE BUD (*Anisocoma acaulis*) has a similar flower—without a red center—and is often found in the same habitat as Desert Dandelion. Its 2 to 4-inch, deeply-lobed, oblong, light green leaves are basal. The involucre bracts are edged with red.

SILVER PUFFS (*Microseris lindleyi*), also, has a head consisting of only ray flowers like Desert Dandelion, Chicory, and Scale Bud. Its flower is smaller but the involucre bracts are up to an inch high and several protrude out beyond the rays. The seed head is very attractive—a silver puffball. Bloom is from April to June below 6,000 feet elevation.

YELLOW TACK-STEM (*Calycoseris parryi*) is easily identified by the many tack-shaped glands covering the stems. The yellow ray flowers are delicately lobed at the tips and have a brownish-purple stripe on the underside. It is scattered throughout the Scenic Area and blooms in April and May.

Desert Dandelion

Scale Bud

Yellow Tack-Stem

Silver Puffs

Silver Puffs seed head

DESERT MARIGOLD
Baileya multiradiata
Flower: Yellow

Sunflower Family - *Asteraceae*

The bright yellow flowers of Desert Marigold, blooming on roadsides throughout the East Mojave, are real eye-catchers. Gray woolly leaves grow low to the ground. The many flower stems are often over a foot tall and each bears a single inflorescence of ray and disk flowers, 1 to 2 inches across. The perennial herbaceous plants are very hardy and their bloom is persistent from April through summer.

Kelso-Cima Road, between Cima and Cedar Canyon Road, dances with Desert Marigolds all summer.

A similar plant, **WOOLLY MARIGOLD** (*Baileya pleniradiata*), can be found at the base of Kelso Dunes, along Cima Road near I-15, and scattered throughout the East Mojave. It has smaller flowers which bloom atop woolly, leafy, branched stems.

Desert Marigold

DESERT TRUMPET
Eriogonum inflatum ssp. inflatum
Flower: Yellow

Buckwheat Family - *Polygonaceae*

Here is an easily recognized buckwheat with an unusual growth form. It has oval green leaves, about 1 inch broad, in a basal rosette. Erect, leafless stems grow from the rosette, sending out thin branches. The main stem inflates just below the branches. The slender upper branches bear tiny yellow flowers.

Some Desert Trumpets grow up to 4 feet tall and also have inflated upper stems that send out their own slender branches.

Desert Trumpet is a perennial. The dried "trumpet" from the previous year is often still standing when spring growth of new basal leaves begins.

Look for this plant on the Cedar Canyon Wash walk. It is common throughout the Scenic Area.

New and old stems of Desert Trumpet

GOLDEN GILIA—DESERT GOLD
Linanthus aureus var. aureus
Flower: Yellow

Phlox Family - *Polemoniaceae*

Aureaus is Latin for "golden." This annual is like a tiny drop of gold on the desert floor. The plant is not more than 4 inches high. Linear leaves grow in rings around the slender stem. Flowers are funnel-shaped with the corolla forming 5 distinct lobes. The throat has a brown or purplish spot in the center. Bloom begins as early as March.

Look for Golden Gilia along Kelbaker Road east of Baker and in the high country near Rock Springs.

GROUND CHERRY
Physalis spp.
Flower: Yellow

Nightshade Family - *Solanaceae*

Two species of Ground Cherry are common in the East Mojave. Both are perennial and have yellow saucer-shaped corollas. In the high country you will find **IVY-LEAVED GROUND CHERRY** (*Physalis hederaefolia*), and below 4,000 feet, **THICK-LEAVED GROUND CHERRY** (*Physalis crassifolia*).

On these plants the calyx swells to an oval, membranous bag—like a paper lantern—containing the fruit which is an edible berry with many seeds, somewhat like a tiny tomato.

Thick-leaved Ground Cherry grows along the Mojave Road leading to Ft. Piute. Bloom is from March to May. Ivy-leaved Ground Cherry grows almost weed-like under Junipers near Government Holes and blooms from May to July.

Ivy-Leaved Ground Cherry

GROUNDSEL
Senecio multilobatus
Flower: Yellow

Sunflower Family - *Asteraceae*

Here is another plant of the large Senecio genus. This groundsel has a perennial taproot from which new growth appears in spring. The plant is usually 1 to 2 feet high. Leaves are mostly basal, about 3 inches long and deeply lobed.

Bright yellow blooms have 10 to 13 rays and many disks. The involucre at the base of each flower head is made up of long equal-sized bracts. Bloom is in April and May.

Groundsel grows at the higher elevations of the Scenic Area on slopes and in washes. You'll probably find some behind Rock Springs and in Caruthers Canyon.

LITTLE GOLD POPPY
Eschscholtzia minutiflora
Flower: Yellow

Poppy Family - *Papaveraceae*

Here is a miniature jewel, sparkling on the desert floor. The 4-petaled flowers of this tiny annual grow on short peduncles along the stems above fine, lacy leaves. Sepals are joined to form a cap which falls off when the flower opens. This is a characteristic of all poppies. Fruit is a tapering capsule filled with many seeds.

Little Gold Poppy blooms in early spring and is common along roadsides and on the gravely banks of washes, usually up to 5,000 feet elevation.

A similar, larger-flowered poppy is also found in the East Mojave. **DESERT GOLD POPPY** (*Eschscholtzia glyptosperma*) sometimes carpets portions of Lanfair Valley a brilliant yellow.

Little Gold Poppy

Desert Gold Poppy

LIVE-FOREVER
Dudleya saxosa
Flower: Yellow

Stonecrop Family - *Crassulaceae*

A charming little succulent can be found in rocky niches along the creekbed in Caruthers Canyon. Thick fleshy leaves, up to 4 inches long, form a soft green, basal rosette. Pinkish flower stalks are up to 8 inches high topped with delicate yellow, nodding flowers set in reddish calyx lobes.

You may also see Live-forever on the Mary Beal Nature Study Trail and throughout the Providence Mountains. The local name is Hen and Chickens.

RIGID SPINY-HERB—SPINY CHORIZANTHE
Chorizanthe rigida
Flower: Yellow

Buckwheat Family - *Polygonaceae*

In spring this small annual gives some hint of what it will become when it dries out in fall. When the plant shrivels up it is densely spiny all over, like a tiny cactus, and may remain in place for up to a year.

Spiny bracts almost conceal very tiny yellow flowers. The lower leaves are broadly oval on long stems and hug close to the ground. The center of the 2 or 3-inch-high plant is a mass of green flexible spines. It grows under very harsh, hot conditions.

You'll find Rigid Spiny-Herb on the Kelbaker Volcanic Rock Hill Walk growing among the black lava rocks, and near Kelso Dunes.

ROCK PEA—DEER VETCH
Lotus rigidus
Flower: Yellow

Pea Family - *Fabaceae*

Deep lemon-colored "pea" flowers bloom atop slender stems on this 1 to 3-foot-tall perennial plant. Old blossoms fade to red. The small, 1/2 inch, oblong leaves are arranged sparsely along the stems in groups of 3 to 5. You'll see that the calyx and leaves are hairy when viewed through a hand lens. This plant dies back in winter.

Rock Pea can be found on the Cedar Canyon Wash Walk, on the hillsides near Rock Springs and near Mitchell Caverns. Bloom is from March to May.

TANSY MUSTARD
Descurainia pinnata
Flower: Yellow

Mustard Family - *Brassicaceae*

This little annual mustard is an attractive lacy plant. Latin for feathered is "pinnata" which refers to the deeply-incised leaves. It grows to be about 20 inches high but can be much smaller. Leaves are hairy, as seen on close inspection.

Flowers are very tiny, clustered together, with 4 petals and 6 stamens. The slender seed pods are about 1/2 inch long and open from below upward.

Tansy Mustard seeds have been used for food and medicine.

The plant can usually be found growing in the shade of larger shrubs such as Creosote Bush or under the wide spread of Junipers. It's an early bloomer, sometimes the first to show its colors in spring.

There are many other members of the Mustard Family growing in the East Mojave. **PEPPER-GRASS** (*Lepidium lasiocarpum*) is another early bloomer with spicy leaves. It has tiny white flowers and small round delicate fruits notched at the top. It likes to grow under the protection of large shrubs or trees and often forms dense colonies under Junipers in the high country.

PALMER BEAD POD (*Lesquerella palmeri*) is an annual about 1 foot tall. Flowers have four clear yellow petals. Fruits are small round "beads." Look for it along the old Mojave Road leading to Ft. Piute.

Flowers and tiny seed pods of Tansy Mustard

WALLACE DAISY—WOOLLY DAISY
Eriophyllum wallacei
Flower: Yellow

Sunflower Family - *Asteraceae*

Meet one of the "little guys" of the desert. You'll have to get down on your hands and knees to study this plucky annual. In times of drought it may have only one flower head, but in years of good rain a single plant may have 10 or more blooms.

The Greek word for wool is "erion" and this plant has gray woolly leaves. The small bright yellow, sunflower-type blossoms—both rays and disks—are about 1/4 inch wide, each with 5 to 10 ray flowers. The stubby rays usually turn slightly downward. The whole plant is from 1 to 4 inches tall.

For such a tiny fellow, it can be quite conspicuous in sheer numbers. It blooms from March to May throughout the Scenic Area.

YELLOW CUPS—SUNDROPS
Camissonia brevipes
Flower: Yellow

Evening Primrose Family - *Onagraceae*

   The Evening Primrose Family is well-represented in the Scenic Area and this is one of its most beautiful members. Bright yellow, 1 inch, cup-like blooms consist of 4 petals. The broad oval leaves grow mainly in a basal rosette. You'll notice red veins on their undersides. This annual plant has several stems growing up to 1-1/2 feet tall. Fruit is a 2 or 3-inch-long narrow capsule. Look for Yellow Cups blooming from March to May along Kelbaker Road.

BIGELOW MIMULUS—MONKEY FLOWER
Mimulus bigelovii
Flower: Reddish-purple

Figwort Family - *Scrophulariaceae*

The low and mid-elevations of the East Mojave are home to this pretty little annual. You're likely to find it blooming in April in and near washes along Kelbaker Road east of Baker. It's a "belly flower," 2 to 8 inches tall, but the flowers, at the tips of stems, are quite large for a small plant. The reddish-purple corolla has 5 lobes and a yellow throat.

Mimulus is a large genus—Munz lists 77 in California. The genus name is from the Latin word for a comic actor—"mimus"—which refers to their smiling corollas.

BROOMRAPE—BURROWEED STRANGLER
Orobanche cooperi
Flower: Purple

Broomrape Family - *Orobanchaceae*

Literally thrusting its way up through desert soil, this strange parasitic plant definitely arouses curiosity. It lacks leaves and chlorophyll and is entirely dependent on the root system of its host plant—usually Burrobush or Cheesebush—for food and water. The 2-lipped, 5-lobed purple corollas are up to an inch long and completely cover the plant stalk which is up to 1 foot high.

Broomrape may be found blooming in April and May in sandy soils near Kelso and on the Volcanic Rock Hill Walk in the washes.

CHIA
Salvia columbariae
Flower: Blue

Mint Family - *Lamiaceae*

Chia (pronounced chee-ah) is a wonderful, aromatic, little annual that has been an important food source for Native Americans and provides sustenance to many forms of wildlife.

This member of the Sage (*Salvia*) genus has the typical "square," 4-sided stems. Deep green, 4-inch-long leaves are opposite, irregularly divided and wrinkled. They grow mainly at the base of the plant.

Blue flowers grow in round clusters at intervals along the stem. In dry years there may be only 1 cluster at the top; in wet years maybe 2 or 3 clusters. Each flower is 2-lipped. The lower lip has a white center with navy blue spots. Below the clusters are purplish bracts with short spines. Bloom occurs from March to June, or later depending on elevation.

Seeds are "ripe" about 2 months later. The seeds swell when placed in water. They are very small and much effort must have been involved to gather them for food. Native Americans roasted and ground them for pinole.

Salvia in Latin means "to save" and refers to the medicinal qualities of the sages.

Chia can be found throughout the Scenic Area.

DESERT HYACINTH—BLUE DICKS
Dichelostemma pulchella
Flower: Blue

Lily Family - *Liliaceae*

Tall, naked, slender stalks bear a cluster of 4 or more delicate blue flowers. The few long, slender basal leaves are grasslike and appear over a month before the flower stalk. They resemble the leaves of Mariposa Lily, a close relative that grows in the same areas. The plants are usually a little over a foot tall, but some grow much taller, especially with the support of small shrubs. There are 6 petal-like sections to each flower. Desert Hyacinth is perennial from a corm (an underground bulblike stem). Bloom is from March to May.

I walk the slopes of Pinto Mountain weekly and had not seen this plant in the area for 4 years. Then one early April, after good fall rains and a very wet March, I was surprised to see hundreds of Desert Hyacinths ready to bloom. Some of the plants that I marked with rock piles to study were dug up and eaten by animals (Carl says the rock piles show the critters where the plants are) but there were plenty left to bloom. Native Americans and early settlers also enjoyed eating the bulbs. Tulip is a local name.

You'll find this plant along Black Canyon Road. It's common in good flower years.

Desert Hyacinth flowers

DESERT LARKSPUR
Delphinium parishii
Flower: Blue

Crowfoot Family - *Ranunculaceae*

During 5 years of drought in the East Mojave I had seen only one Desert Larkspur—and that was almost at the top of Pinto Mountain. Imagine my delight when hundreds began blooming on our property one late spring, just when the Desert Hyacinths were fading.

Even though these plants have perennial root crowns it took the magical heavy rains of March to stimulate growth which had so long been dormant. Desert plants can be very patient.

Larkspur grows to almost 2 feet tall, with one or several stems. The green, deeply divided leaves grow on long petioles. As many as 25 sky-blue flowers bloom along the stalks. They are unusual, intricate blossoms, worth studying with a hand lens.

You may find Larkspur on the Mojave Road-Ft. Piute hike blooming in April and in Caruthers Canyon blooming in June.

GIANT FOUR O'CLOCK
Mirabilis multiflora ssp. pubescens
Flower: Rose-purple

Four O'Clock Family - *Nyctaginaceae*

Giant Four O'Clock can't fail to attract your attention. This showy perennial with deep rose-colored blooms sprawls in a large mat on the desert floor. Mirabilis in Latin means "wonderful," and it is, indeed, wonderful that these plants grace our desert.

Leaves are up to 3 inches long, slightly oval or heartshaped and grow opposite on the stems. They may be tinged purple on the undersides.

What appears to be the flowers are really colorful funnel-shaped calyxes set in cupshaped involucres (a leaflike structure below the flower). The blooms smell sweet and open in the evening (four o'clock), usually closing in late morning. Blooms are 1 or 2 inches long. They appear from April to August.

In winter the plants dry up. The leaves turn almost white and blow like scraps of paper across the land, seemingly quite indestructible. The plants have a thick woody root from which new growth appears in spring.

One of my favorite Giant Four O'Clocks is on Kelso-Cima Road just north of Cedar Canyon Road near the railroad tracks. You'll see other fine specimens in this area.

GOODDING VERBENA
Verbena gooddingii
Flower: Light purple

Vervain Family - *Verbenaceae*

This is a green, hairy-leafed perennial, with delicate little flower clusters, that is one of the earliest to bloom at the higher elevations of the Scenic Area.

Goodding Verbena has several stems, growing up to 1 foot high, often much shorter. The leaves are lobed and divided. Each flower, growing in clusters at the top of the stems, consists of a 5-lobed corolla with a deep-tubed throat. They are about 1/2 inch wide with a 1/2-inch-long tube. The flowers in one cluster usually do not open all at the same time.

Bloom is from March through July with another sometimes in late summer or fall.

This verbena grows on slopes and in canyons especially near Providence, New York, Ivanpah and Clark Mountains.

LOCOWEED—RATTLEPOD—RATTLEWEED—MILKVETCH
Astragalus spp.
Flower: Purple, Blue, White, Red

Pea Family - *Fabaceae*

There are over 400 species of Astragalus in North America, and many grow in the desert, making identification difficult.

It is one of the first bloomers along the roadsides, sending up racemes of usually purple or blue flowers. Each small pealike blossom has a banner (upper larger petal), wings (2 side petals) and keel (2 united lower petals). The leaves are pinnate (leaflets arranged along each side of the stem, feather-like).

In many species the flowers are followed by inflated papery seed pods which dry out and "rattle" when moved.

Many of these plants contain great amounts of selenium which can be poisonous to animals, making them "mad," hence the name Loco-weed, meaning "crazy" in Spanish.

1939 to 1941 were drought years in the East Mojave and the range cattle resorted to eating Locoweed which was about the only green available. Ranchers were forced to ship their cows out of the area. A local rancher told me how you can tell if a cow has been into Locoweed—the hair on its forehead sticks straight out, its step isn't quite right and it doesn't take in water when slurping at a trough. When other forage is available the cattle shun Locoweed.

Two that are fairly easy to identify: **BORREGO LOCOWEED** (*Astragalus lentiginosus borreganus*) is common on and near Kelso Dunes. It is soft and gray with fruit like a small pea pod that does not strongly inflate. **LAYNE LOCOWEED** (*Astragalus layneae*) has hairy, sickle-shaped, purple-tinted pods.

Borrego Locoweed on Kelso Dunes

Layne Locoweed

MOJAVE ASTER—DESERT ASTER
Xylorhiza (Machaeranthera) tortifolia
Flower: Lavender

Sunflower Family - *Asteraceae*

Surely this stunning plant has escaped from someone's prized flower garden. Showy lavender flowers bloom atop 1 to 2-foot-high leafless stems growing from a woody base. Leaves are up to 2 inches long, hairy, with spine-tipped margins. The 2-inch-wide flowers consist of 40 to 60 lavender rays and yellow center disks. Often 20 blooms may be on a single plant.

Mojave Aster is usually found on dry rocky slopes. You'll find it blooming in April and May on the north side of the Volcanic Rock Hill off Kelbaker Road. Look for it, also, in the Granite Mountains north of Interstate 40 and along Black Canyon Road between Hole-in-the-Wall and Essex Road. Depending on rains, these plants may bloom again in fall.

NOTCH-LEAFED PHACELIA—WILD HELIOTROPE
Phacelia crenulata ssp. ambigua
Flower: Purple or Blue

Waterleaf Family - *Hydrophyllaceae*

Don't touch this plant! It's a striking 1 to 2-foot-tall annual with deep purple or blue flowers and shiny leaves and you'll want to examine it closely. But many members of the Phacelia genus cause a terrible itching rash on some people, lasting several weeks.

Dark green leaves, up to 4 inches long, are deeply lobed and strongly scented—unpleasantly so to my sense of smell. Bell-shaped, 1-inch-long flowers bloom in terminal clusters. Phakelos—from which we get the word Phacelia—is Greek for "cluster."

One morning in early May I hiked the old Mojave Road to Ft. Piute. The strong scent of these Phacelias, growing in thick masses on the trail, became overpoweringly nauseating.

I came home with beautiful pictures of this plant along with a not-so-nice souvenir—a bad rash.

Another Phacelia you may meet in the Scenic Area is **CANTERBURY BELL** (*Phacelia campanularia*). This handsome annual is often cultivated in gardens although it, also, may cause severe skin problems.

The bell-shaped flowers are deep blue, about 1 inch long. The plant may have several stems, up to 2 feet tall, with broad leaves. It prefers sandy, gravelly soils below 4,000 feet elevation.

I found Canterbury Bell blooming in April on the Mary Beal Nature Study Trail at Mitchell Caverns.

Notch-Leaved Phacelia

PIÑON ASTER
Machaeranthera canescens
Flower: Light purple

Sunflower Family - *Asteraceae*

    The welcome bloom of this short-lived perennial occurs in summer through October when most other flowers have faded. The plants are 2 to 3 feet high with several branches and have small, linear leaves. Blooms have 8 to 25 pale purple rays surrounding a yellow center.

    This aster grows in the Piñon-Juniper belt above 4,500 feet elevation in the Scenic Area.

PURPLE MAT
Nama demissum
Flower: Reddish-purple

Waterleaf Family - *Hydrophyllaceae*

Here is another "belly" flower that requires a hands and knees position to really study the bright little flowers. The plant is only 2 or 3 inches high. The stems of this spring blooming annual spread prostrately along the ground. The narrow green leaves are clustered, up to 3/4 inch long. Corollas have 5 round lobes. Look for Purple Mat along Kelbaker Road. You may come across it on the Volcanic Rock Hill Walk in the nearby washes.

WINDMILLS
Allionia incarnata
Flower: Magenta

Four O'Clock Family - *Nyctaginaceae*

What appears to be a single magenta bloom on this sprawling, ground-hugging plant is actually 3 flowers growing together. If you gently pull on the petals you will see that each is a separate flower with its own stamens and pistil.

The prostrate hairy stems are sticky, often covered with sand and dirt. This is a perennial that dies back in winter, sending up new growth in early spring.

Bloom usually begins in April. Look for Windmills on the desert flats near Kelso and on the Mojave Road-Ft. Piute hike. There are many near Bonanza King Mine in the Providence Mountains.

YELLOW THROATS
Phacelia fremontii
Flower: Light purple

Waterleaf Family - *Hydrophyllaceae*

If you bend down and sniff this little annual, you'll be surprised. It smells just like a skunk!

The leaves grow at the base of the plant and are deeply lobed, 1 or 2 inches long. The small 5-lobed corolla is blue to lavender with a yellow throat.

This Phacelia is common in the Scenic Area and often grows in such profusion the roadsides appear light purple. The sandy banks along Morningstar Mine Road are a good spot to look for Yellow Throats blooming in April.

COOPER DYSSODIA
Dyssodia cooperi
Flower: Orange

Sunflower Family - *Asteraceae*

Stinky is the best word to describe Dyssodia, which in Greek means "a disagreeable smell." Cooper Dyssodia is perennial from a woody base, about 12 to 18 inches tall, with many stems covered with tiny hairs. Small leaves are linear with spine-tipped margins. The 8 to 12 yellowish-orange ray flowers are thin and much lower than the disk flowers. The involucre bracts have amber gland dots. Bloom is in May and June and again in fall.

Dyssodia is common at elevations between 2,000 and 5,000 feet. It lines Kelbaker Road and Kelso-Cima Road.

DESERT FIVE SPOT
Eremalche rotundifolia
Flower: Pink

Mallow Family - *Malvaceae*

The leaves on this charming annual are almost as attractive as its unusual bloom. Rotundifolia means "round leaved."

Under favorable conditions the plant grows over a foot tall. It may have several hairy branches. The few rounded, toothed leaves are up to 2 inches wide. The globe-like pink corollas do not open wide. Each of the 5 overlapping petals has a red spot at the base. You have to peek inside from the top to see all 5 spots. This plant prefers elevations below 4,000 feet.

You may find Desert Five Spot on Essex Road leading to Mitchell Caverns or on the Volcanic Rock Hill Walk where I found several growing among the black rock right on top of the hill. Bloom is from March to May.

Leaves of Desert Five Spot

DESERT MALLOW—APRICOT MALLOW—SORE EYE POPPY—GLOBE MALLOW—DESERT HOLLYHOCK
Sphaeralcea ambigua
Flower: Orange

Mallow Family - *Malvaceae*

Most desert travelers are familiar with this rather ubiquitous woody-based perennial. But no matter how often I see it, I'm impressed with the sculptured beauty of its blooms.

Desert Mallow is usually between 1 and 3 feet tall. Leaves are up to 2 inches long, with scalloped edges and appear very wrinkled. All parts of the plant except the blooms are covered with tiny hairs which are irritating to some people (sore-eyes). If you have a hand lens be sure to study the leaves—they look like they're covered with cactus spines.

The 5 petals overlap to form a rounded, cup-shaped bloom. Flowers appear from March through June.

It grows on rocky slopes, in canyons, and along disturbed roadsides up to 5,000 feet elevation. There is usually a good Mallow display on Kelso-Cima Road heading south from Cima.

EATON FIRECRACKER—FIRECRACKER PENSTEMON
Penstemon eatonii
Flower: Red

Figwort Family - *Scrophulariaceae*

Here is a plant that, when in bloom, really catches your eye. The tubular flowers are bright scarlet, growing on the sides of long stems.

The Penstemon genus has opposite leaves, a 5-part calyx, tubular corollas, 4 fertile stamens, and a 5th sterile stamen (pente means "five", stemon means "stamen" in Latin) that is often "bearded." Penstemons are sometimes called Beard-tongues.

Eaton Firecracker has perennial roots. The large green leaves, often with a purple cast, are opposite on the stems. The plant grows to about 3 feet.

Hummingbirds feed on the nectar. Many long tube-shaped flowers, like Eaton Firecracker, are pollinated by hummingbirds.

It is common but widely scattered. Look for it in Caruthers Canyon or areas surrounding New York and Clark Mountains.

FIDDLENECK
Amsinckia tessellata
Flower: Orange

Borage Family - *Boraginaceae*

Hairy is the word to describe this bristly common annual. The plant is usually about 1 foot tall, most of it consisting of leaves. Tiny tubular yellow-orange flowers grow in coils much like those that top the necks of violins. Fruit is a 4-part nutlet. Bloom begins in late March.

There are great numbers of Fiddlenecks growing in the shade of Creosote Bushes along Kelbaker Road and beneath the spread of Junipers in the high country.

Other common members of the Borage Family that grow by the hundreds in the Scenic Area are **FORGET-ME-NOTS** (*Cryptantha spp.*) These plants are tricky to correctly identify until they have fruit, by which time we may have lost interest. They are usually hairy, with veinless leaves, and have tiny, white, 5-lobed flowers, often growing in coils like Fiddleneck.

An interesting white-flowered, close relative is **ARIZONA POP-CORN FLOWER** (*Plagiobothrys arizonicus*) with leaves which have purple midribs and margins. If you break a stem, red juice oozes out. Sometimes my shoes have red spots after walking in a field filled with Popcorn Flower. A local rancher told me his cattle favor this plant and their mouths are often stained red. He calls it Blood Weed. They are very common in the Government Holes-Rock Springs area and near Hole-in-the-Wall.

Fiddleneck

FILAREE—HERON BILL—CLOCKS—STORKSBILL
Erodium cicutarium
Flower: Pink

Geranium Family - *Geraniaceae*

The tiny rosettes of Filaree leaves poke out of February snows and some flowers have already gone to seed by early March. There are several species of Filaree in the West. As a child I made "scissors" with the long tapering seed capsules—threading one point into the bottom fat portion of another.

The bloom of this desert species has 5 sepals, 5 petals, 5 stamens, and varies from light pink to a darker purple. The leaves are green, tinged with red.

The most easily identified part of the plant is the fruit. It is 5-lobed at the base and contains 5 seeds. It's about 1 or 2 inches long, tapering at the top to needle size. When dry, this seed head falls off and curls and uncurls with variations in humidity, thus enabling it to penetrate the soil with its seeds to begin life anew.

Heron in Greek is "erodios," and it was thought that the seed capsule resembled a heron's bill.

Filaree has been used medicinally and gastronomically by Native Americans and settlers. It was introduced from the Mediterranean early in our country's history. It's a favorite of cattle and contains 26% protein so is one of the most-prized forage plants of the ranchers.

The plant is common along roadsides and in disturbed areas. It is usually small but in years of heavy rainfall it may grow a foot tall.

INDIAN PAINTBRUSH—DESERT PAINTBRUSH
Castilleja chromosa
Flower: Red

Figwort Family - *Scrophulariaceae*

That flash of red you see out your car window from early spring through summer is probably Indian Paintbrush.

There are over 30 species of Castilleja in California. The genus name honors a Spanish botanist. Chroma in Greek means "color." The vivid red color of the plants is not from the flower, for it is completely hidden from view. The bracts—reduced leaves below the flower—and the calyxes—the sepals—are the colorful parts, much like the cultivated bougainvillea of the Southwest.

The plant is perennial from an underground rootstalk. It sends up thin reddish-green leaves in February or early March and is in bloom by April. Castillejas are partially parasitic on the roots of other plants.

A usually reliable patch of Indian Paintbrush can be found on the north side of Cedar Canyon Road across from the Rock House above Rock Springs. You'll find it, also, on the Teutonia Peak Trail.

LONGLEAF PHLOX
Phlox viridis ssp. compacta
Flower: Pink or White

Phlox Family - *Polemoniaceae*

High elevations are home to this pretty little low-growing plant. It is perennial, from a woody root crown, dying back in winter. The narrow leaves, about 1 inch long and hairy, grow alternately on slender reddish stems. The whole plant is only 5 to 8 inches high. The pink or white flowers have long tubes and 5 distinct lobes. Bloom begins in April.

Phlox prefers rocky hillsides near New York, Clark and Providence Mountains about 4,000 to 5,000 feet elevation. It is not too common, but a joy to find.

MARIPOSA LILY—DESERT MARIPOSA
Calochortus kennedyi var. kennedyi
Flower: Orange

Lily Family - *Liliaceae*

It's hard to believe this small bright flower is related to the edible garden asparagus, but the two plants have those qualities in common to fit into the Lily Family. Both are perennials.

After the flower and fruit cycle is complete, the Mariposa Lily dries up and blows away. A large underground bulb (corm) is its food storage chamber for the next growing season.

In early spring thick, green, grass-like leaves, some curled in corkscrew shapes, emerge in rocky soil—one to a plant. They are readily eaten by small animals so what seems to promise to be a good bloom often results in just a few plants going through their entire cycle.

Flowers appear from April to June. There are 3 fan-shaped, usually overlapping, orange petals, each with a hairy purple patch at the base. If the plant is growing out in the open it is quite short but if it grows up through another bush it can be almost 2 feet tall. In the higher mountains of the East Mojave you may find a flower with yellow petals (*var. munzii*). Fruit is a long, angled capsule with vertical white stripes.

The bulb of the Mariposa Lily was eaten by Native Americans.

Calochortus is Greek for "beautiful grass." The Spanish word for butterfly is "mariposa."

Mariposa Lilies are common and can be found on the low hillsides and desert flats around Rock Springs and Government Holes and in Lanfair Valley.

Flower and seed pod of Mariposa Lily

SAND VERBENA
Abronia villosa
Flower: Pink

Four O'Clock Family - *Nyctaginaceae*

Even though Sand Verbena is familiar to most desert visitors, it is a great favorite, and the large trailing mats covered with bright pink, fragrant flowers are eagerly awaited.

The opposite leaves, up to 1-1/2 inches long, are hairy and sticky so that grains of sand adhere to them.

Flower heads are made up of clusters blooming from stalks that grow from the leaf axils. The heads are 2 or 3 inches wide.

Each flower making up the head has a 5-lobed corolla with a slender throat. The fruit has 3 to 5 wings.

Sand Verbena blooms from February to mid-summer. Look for the colorful low-growing mats along Kelbaker Road east of Baker, near Kelso Dunes, and along Essex Road leading to Mitchell Caverns.

SPANISH NEEDLE
Palafoxia arida var. arida
Flower: Pink

Sunflower Family - *Asteraceae*

This hardy annual tolerates harsh conditions, growing in alkali soils and in sand, usually below 2,500 feet elevation.

It is an erect, open-branching plant, 1 to 2 feet tall. The greenish-gray leaves are about 2 inches long. Flowering flat-topped heads are made up of 10 to 20 tubular disk flowers. Bloom begins in early spring.

Look for Spanish Needle along sandy roadsides in the low elevations of the East Mojave. It grows at the base of Kelso Dunes. In June I found several plants in bloom on the road to Bonanza King Mine and in July they were still flowering along Kelbaker Road north of I-40.

WILD RHUBARB—CANAIGRE—DOCK
Rumex hymenosepalus
Flower: Pink

Buckwheat Family - *Polygonaceae*

Yes, the stalks of this plant can be used as a rhubarb substitute if properly prepared to remove the tannin.

You are likely to see Wild Rhubarb along disturbed roadsides up to 5,000 feet elevation. It's common near Kelso, Hole-in-the-Wall and Camp Rock Springs. The leaves may be 10 inches long, with wavy margins. Stems resemble rhubarb with a pinkish cast. The plants are about 2 feet high—some almost twice that size. During times of drought they may be only a few inches tall.

Clusters of papery, pink fruits are much showier than the blooms.

Wild Rhubarb is perennial from tuberous roots. The roots and leaves have a high tannin content and have been used for tanning leather.

INTRODUCED LANDSCAPE PLANTS

Two non-native, drought-resistant plants are commonly used for landscaping in the East Mojave. **BIRD OF PARADISE** or **PARADISE POINCIANA** (*Caesalpinia gilliesii*), introduced from South America, is in the Pea Family. It has showy yellow flowers with 4 or 5-inch-long red stamens. These plants bloom all summer. A row of these shrubs grows in front of the Cima Post Office/Store.

The deciduous **CHINABERRY TREE** (*Melia azedarach*) from Asia is in the Mahogany Family. It provides dense shade under its dome-shaped crown of deep green leaves. Clusters of lilac-colored flowers are followed by hard, yellow, berrylike fruits which are poisonous.

Bird of Paradise at Cima Store

RESOURCES

BUREAU OF LAND MANAGEMENT OFFICES

California Desert District
6221 Box Springs Blvd., Riverside, CA 92507
714 / 653-6950

California Desert Information Center
831 Barstow Rd., Barstow, CA 92311
619 / 256-8617

Barstow Resource Area
150 Coolwater Lane, Barstow, CA 92311
619 / 256-3591

Needles Resource Area
101 W. Spikes Rd., Needles, CA 92363
619 / 326-3896

SOURCES FOR NATIVE PLANTS AND SEEDS

The Theodore Payne Foundation
10459 Tuxford St., Sun Valley, CA 91352
818 / 768-1802

Plants of the Southwest
930 Baca St., Santa Fe, NM 87501
505 / 983-1548
(Mail order shrubs, trees, seeds)

Cactus Mart
49889 State Highway 62, Morongo Valley, CA
(Open daily 10 a.m. to 6 p.m.)

Living Desert
47900 Portola Avenue, Palm Desert, CA 92260
619 / 346-5694
(Desert gardens, animals, birds, plant nursery. Open daily 9 a.m. to 6 p.m.—closed June 15 through August 31)

Rancho Santa Ana Botanic Garden
1500 N. College Ave., Claremont, CA 91711
(Native gardens, annual plant sales)

MAPS

"A Recreational Guide to the East Mojave National Scenic Area" available for $3.50 to Desert Protective Council Foundation, PO Box 76210, Los Angeles, CA 90076.

Automobile Club of Southern California map of San Bernardino County.

Local Bureau of Land Management offices have area topographical maps available.

ORGANIZATIONS

Desert Studies Center (Zzyzx Road off I-15)
c/o Biology Dept, California State University
Fullerton, CA 92634
714 / 773-2428

A University Consortium operates this field station for research and education. The Barstow BLM Resource Area leads tours on weekends from Oct. to April. Call them at 619 / 256-3591 for information. The Center sells books of local interest and has a small gift shop.

An East Mojave Volunteer Association is working to assist The Needles Resource Area Bureau of Land Management in their programs.

Friends of the Mojave Road is located in the historic Goffs School House. Dennis Casebier, Chairman. HCR G, No. 15, Essex, CA 92332. 619 / 733-4482

Guidebooks to 4-wheel-drive roads; books on the history of the East Mojave.

Southern California Botanists issues a journal, "Crossosoma," that often deals with desert plants. They sponsor trips to many desert locations throughout the year and hold annual plant sales. Contact them at Rancho Santa Ana Botanic Garden, 1500 N. College Ave., Claremont, CA 91711.

California Native Plant Society publishes "Fremontia" which has articles on plants, nursery advertisements and book reviews. Contact them at 909 12th St. #116, Sacramento, CA 95814.

GLOSSARY

achene - small, dry one-seeded fruit.

alternate - placed singly, at different heights in succession, on each side of a stem.

annual - a plant that grows from seed, flowers, produces seeds and dies in one season.

areole - a small space on cacti from which spines and glochids grow.

awn - a terminal slender bristle.

axil - the angle between a leaf and its supporting stem.

biennial - a plant that takes two years from seed to death.

bract - a reduced leaf structure beneath a flower.

bulb - an underground leaf bud with thick scales.

calyx - all of the sepals together.

catkin - a dangling flower spike.

corm - the enlarged, fleshy, underground base of a stem.

corolla - all of the petals together.

deciduous - falling off at maturity, in winter, or during drought.

disk flowers - tubular flowers in the heads of Asteraceae Family as distinct from ray flowers.

drupe - a fleshy fruit with a hard shell enclosing one seed (peach, cherry).

endemic - natural to a specific place and not found elsewhere.

evergreen - a plant that keeps its leaves throughout the entire year.

family - the major subdivision of an order in the classification of plants.

fascicled - clustered or bundled in a bunch.

fruit - the developed, ripened pistil and all of its parts.

genus - the major subdivision of a plant family (pl. genera).

glochids - the short hairs or bristles on cacti different from spines.

grass - plants having jointed stems, sheathed leaves, and seed-like grains.

head - a dense cluster of flowers arising from the same point, without stems.

herb - a plant without woody stems.

inflorescence - a cluster of flowers. The arrangement of flowers on the axis.

involucre - the bracts beneath a flower cluster.

linear - long, thin and narrow.

node - the joint of a stem where a leaf or branch is inserted.

opposite - one on each side of the stem—having two leaves at one node.

palmate - handshaped with fingers spread; radiating from a common point.

peduncle - the stem of a flower cluster.

perennial - having a life cycle lasting for many years.

petal - one of the segments of a corolla.

petiole - the stalk of a leaf.

pinnate - like a feather. Leaflets arranged along each side of a common petiole.

pistil - the female part of a flower which will mature into fruit, consisting of stigma, style and ovary.

ray flowers - the marginal florets in flower heads of the Asteraceae Family as distinct from disk flowers.

rhizome - an underground stem or rootstock producing leafy shoots on the upper side and roots on the lower side.

rosette - a circular cluster of radiating leaves.

sepal - the leafy portion of a flower, often green, beneath the petals.

shrub - a woody plant, smaller than a tree, usually with several stems arising from the base.

sp., spp. - abbreviation for species (singular and plural).

species - the major subdivision of a genus composed of related plants which are able to breed among themselves but not with members of another species.

spikelet - a grass flower cluster.

ssp. - abbreviation for subspecies.

stamen - the male portion of the flower containing the pollen, consisting of a filament and anther.

stigma - the top portion of the pistil which receives the pollen.

style - the mid part of the pistil which connects stigma and ovary.

var. - abbreviation for variety.

whorl - a circle around the same point on an axis.

REFERENCES
SUGGESTED READING

Abrams, Leroy. *Illustrated Flora of the Pacific States* in four volumes. Stanford: Stanford University Press, 1940.

Benson, Lyman. *The Native Cacti of California.* Stanford: Stanford University Press, 1969.

Bowers, Janice Emily. *Seasons of the Wind.* Flagstaff, Ariz.: Northland Press, 1986.

Clarke, Charlotte Bringle. *Edible and Useful Plants of California.* Berkeley: University of California Press, 1977.

Collins, Barbara J. *Key to Trees and Shrubs of the Deserts of Southern California.* Thousand Oaks: California Lutheran College, 1976.

Crampton, Beecher. *Grasses in California.* Berkeley: University of California Press, 1974.

Crittenden, Mabel. *Trees of the West.* Millbrae, Calif: Celestial Arts, 1977.

Dawson, E. Yale. *The Cacti of California.* Berkeley: University of California Press, 1982.

Dodge, Natt N. *Flowers of the Southwest Deserts.* Globe, Ariz.: Southwestern Monuments Assoc., 1965.

Dodge, Natt N. *100 Desert Wildflowers in Natural Color.* Tucson, Ariz.: Southwestern Monuments Assoc.

Dole, Jim W. and Rose, Betty B. *Shrubs and Trees of the Southern California Deserts.* Sepulveda, Calif.: Footloose Press, 1991.

Elmore, Frances H. *Shrubs and Trees of the Southwest Uplands.* Globe, Ariz.: Southwest Parks and Monuments Assoc., 1976.

Harrington, H. D. *How to Identify Plants.* Athens, Ohio: Swallow Press, 1981.

Hitchcock, A. S. *Manual of the Grasses of the United States.* 2 volumes. New York: Dover Publications, Inc., 1971

Jackson, Earl. *Flowering Plants of the Lake Mead Region.* Globe, Ariz.: Southwest Parks and Monuments Assoc.

Jaeger, Edmund C. *Desert Wild Flowers.* Stanford: Stanford University Press, 1972.

MacMahon, James A. *Deserts.* New York: Alfred A. Knopf, 1987.

Mockel, Henry R. and Beverly. *Mockel's Desert Flower Notebook.* Twentynine Palms, Calif., 1971.

Moore, Michael. *Medicinal Plants of the Mountain West.* Santa Fe: Museum of New Mexico, 1979.

Munz, Philip A. *California Desert Wildflowers.* Berkeley: University of California Press, 1962.

Munz, Philip A. *A Flora of Southern California.* Berkeley: University of California Press, 1974.

Ornduff, Robert. *Introduction to California Plant Life.* Berkeley: University of California Press, 1974.

Patraw, Pauline M. *Flowers of the Southwest Mesas.* Globe, Ariz.: Southwest Parks and Monuments Assoc., 1977.

Peterson, P. Victor. *Native Trees of Southern California.* Berkeley: University of California Press, 1966.

Pomeroy, Flora. *Granite Mountain Spring, An Introduction to the Eastern Mojave Desert, California.* Santa Cruz: Environmental Field Program, University of California, 1986.

Raven, Peter H. *Native Shrubs of Southern California.* Berkeley: University of California Press, 1982.

Schad, Jerry. *California Deserts.* Helena, Mont.: Falcon Press Publishing Co., 1988.

Smith, Jr. James Payne and Berg, Ken. *Inventory of Rare and Endangered Vascular Plants of California.* Sacramento: California Native Plant Society, 1988.

Spellenberg, Richard. *The Audubon Society Field Guide to North American Wildflowers - Western Region.* New York: Alfred A. Knopf, 1988.

Stein, Bruce A. and Warrick, Sheridan F. *Granite Mountains Resource Survey.* Santa Cruz: Environmental Field Program, University of California, 1979.

Sudworth, George B. *Forest Trees of the Pacific Slope.* Washington, DC: Government Printing Office, 1908.

Thorne, Robert F. and Prigge, Barry A. and Henrickson, James. "A Flora of the Higher Ranges and the Kelso Dunes of the Eastern Mojave Desert in California" in *Aliso.* Claremont, Calif.: Rancho Santa Ana Botanic Garden, Vol 10, No 1, Oct 1981.

Venning, Frank D. *A Guide to Field Identification - Wildflowers of North America*. New York: Golden Press, 1984.

Watts, Mary Theilgaard and Tom. *Desert Tree Finder*. Berkeley, Calif.: Nature Study Guide Publishers, 1974.

INDEX BY FAMILIES

AGAVACEAE - AGAVE
Agave deserti - Desert Agave
Agave utahensis nevadensis - Pygmy Agave
Yucca baccata - Blue Yucca
Yucca brevifolia - Joshua Tree
Yucca shidigera - Mojave Yucca

ANACARDIACEAE - SUMAC
Rhus trilobata - Squawbush

ASCLEPIADACEAE - MILKWEED
Asclepias erosa - Desert Milkweed

ASTERACEAE - SUNFLOWER
Acamptopappus sphaerocephalus - Goldenhead
Ambrosia dumosa - Burrobush
Ambrosia eriocentra - Wooly-Fruited Bursage
Anisocoma acaulis - Scale Bud
Artemisia ludoviciana - Piñon Wormwood
Artemisia tridentata - Big Sagebrush
Baccharis glutinosa - Water Wally
Baccharis sergiloides - Squaw Waterweed
Baileya multiradiata - Desert Marigold
Baileya pleniradiata - Woolly Marigold
Brickellia californica - California Brickellia
Brickellia incana - Woolly Brickellia
Brickellia oblongifolia linifolia - Piñon Brickellia
Calycoseris parryi - Yellow Tack Stem
Chaenactis fremontii - Fremont Pincushion
Chrysothamnus nauseosus - Golden Rabbitbrush
Chrysothamnus paniculatus - Black-Banded Rabbitbrush
Cirsium neomexicanum - New Mexico Thistle
Dyssodia cooperi - Cooper Dyssodia
Encelia frutescens - Rayless Encelia
Encelia virginensis - Virgin River Encelia
Ericameria cooperi - Cooper Goldenbush
Ericameria laricifolia - Turpentine Brush
Ericameria linearifolia - Linear-Leaved Goldenbush
Erigeron pumilus concinnoides - Tidy Fleabane
Eriophyllum wallacei - Wallace Daisy
Gutierrezia microcephala - Matchweed
Hymenoclea salsola - Cheesebush
Layia glandulosa - White Tidy Tips
Leucelene ericoides - White Aster
Machaeranthera canescens - Piñon Aster
Malacothrix glabrata - Desert Dandelion
Microseris lindleyi - Silver Puffs
Palafoxia arida - Spanish Needle
Pectis papposa - Chinch Weed
Pluchea sericea - Arrowweed
Psilostrophe cooperi - Paperflower
Rafinesquia neomexicana - Desert Chicory
Senecio douglasii - Sandwash Groundsel
Senecio multilobatus - Groundsel
Tetradymia stenolepis - Mojave Horsebrush
Viguiera deltoidea parishii - Goldeneye
Xylorhiza tortifolia - Mojave Aster

BERBERIDACEAE - BARBERRY
Mahonia haematocarpa - Barberry

BIGNONIACEAE - BIGNONIA
Chilopsis linearis - Desert Willow

BORAGINACEAE - BORAGE
Amsinckia tessellata - Fiddleneck
Cryptantha spp. - Forget-Me-Not
Plagiobothrys arizonicus - Arizona Popcorn Flower

BRASSICACEAE - MUSTARD
Descurainia pinnata - Tansy Mustard
Dithyrea californica - Spectacle Pod
Lepidium fremontii - Desert Alyssium
Lepidium lasiocarpum - Pepper Grass
Lesquerella palmeri - Palmer Bead Pod
Stanleya pinnata - Prince's Plume

CACTACEAE - CACTUS
Coryphantha vivapara deserti - Pincushion Cactus
Echinocactus polycephalus - Cottontop Cactus

Echinocereus engelmannii - Hedgehog Cactus
Echinocereus triglochidiatus - Mojave Mound Cactus
Ferocactus acanthodes - Barrel Cactus
Mammillaria tetrancistra - Fishhook Cactus
Opuntia acanthocarpa - Buckhorn Cholla
Opuntia basilaris - Beavertail Cactus
Opuntia bigelovii - Teddy Bear Cholla
Opuntia chlorotica - Pancake Cactus
Opuntia echinocarpa - Silver Cholla
Opuntia erinacea - Old Man Cactus
Opuntia erinacea ursina - Grizzly Bear Cactus
Opuntia phaeacantha - Mojave Prickly Pear
Opuntia ramosissima - Pencil Cholla

CAPPARIDACEAE - CAPER
Cleome isomeris - Bladderpod

CHENOPODIACEAE - PIGWEED
Atriplex canescens - Four-Wing Saltbush
Atriplex hymenelytra - Desert Holly
Atriplex polycarpa - Allscale
Ceratoides lanatum - Winter Fat
Grayia spinosa - Spiny Hopsage
Salsola iberica - Tumbleweed

CRASSULACEAE - STONECROP
Dudleya saxosa - Live Forever

CUCURBITACEAE - GOURD
Cucurbita palmata - Coyote Melon

CUPRESSACEAE - CYPRESS
Juniperus californica
Juniperus osteosperma

CUSCUTACEAE - DODDER
Cuscuta spp. - Dodder

EPHEDRACEAE - JOINT FIR
Ephedra nevadensis - Mormon Tea
Ephedra viridis - Mormon Tea

EUPHORBIACEAE - SPURGE
Chamaesyce albomarginata - Rattlesnake Weed
Croton californicus - California Croton

FABACEAE - PEA
Acacia greggii - Catclaw
Astragalus layneae - Layne Locoweed
Astragalus lentiginosus borreganus - Borrego Locoweed
Caesalpinia gilliesii - Bird of Paradise
Cassia armata - Desert Senna
Krameria grayi - White Ratany
Krameria parvifolia - Range Ratany
Lotus rigidus - Rock Pea
Prosopis glandulosa - Mesquite
Psorothamnus fremontii - Indigo Bush
Psorothamnus spinosus - Smoke Tree

FAGACEAE - BEECH
Querqus turbinella - Turbinella Oak

GERANIACEAE - GERANIUM
Erodium cicutarium - Filaree

HYDROPHYLLACEAE - WATERLEAF
Nama demissum - Purple Mat
Phacelia campanularia - Canterbury Bell
Phacelia crenulata ambigua - Notch-Leafed Phacelia
Phacelia fremontii - Yellow Throats

KRAMERIACEAE - KRAMERIA
Krameria parvifolia imparata - Range Ratany
Krameria grayi - White Ratany

LAMIACEAE - MINT
Salazaria mexicana - Paperbag Bush
Salvia columbariae - Chia
Salvia dorrii - Blue Sage
Salvia mojavensis - Mojave Sage

LILIACEAE - LILY
Calochortus kennedyi - Mariposa Lily
Calochortus kennedyi munzii - Munz Mariposa Lily
Dichelostemma pulchella - Desert Hyacinth
Hesperocallis undulata - Desert Lily

LOASACEAE - LOASA
Petalonyx thurberi - Sandpaper Plant

LORANTHACEAE - MISTLETOE
Phoradendron californicum - California Mistletoe
Phoradendron juniperinum - Juniper Mistletoe

MALVACEAE - MALLOW
Eremalche rotundifolia - Desert Five Spot
Sphaeralcea ambigua - Desert Mallow

MELIACEAE - MAHOGANY
Melia azedarach - Chinaberry Tree

NYCTAGINACEAE - FOUR O'CLOCK
Abronia villosa - Sand Verbena
Allionia incarnata - Windmills
Mirabilis bigelovii - Wishbone Bush
Mirabilis multiflora - Giant Four O'Clock

OLEACEAE - OLIVE
Forestiera neomexicana - Desert Olive
Menodora spinescens - Spiny Menodora

ONAGRACEAE - EVENING PRIMROSE
Camissonia brevipes - Yellow Cups
Camissonia clavaeformis - Brown-Eyed Primrose
Oenothera avita - White Evening Primrose
Oenothera caespitosa - Desert Primrose
Oenothera deltoides - Dune Primrose

OROBANCHACEAE - BROOMRAPE
Orobanche cooperi - Broomrape

PAPAVERACEAE - POPPY
Argemone corymbosa - Prickly Poppy
Eschscholtzia glyptosperma - Desert Gold Poppy
Eschscholtzia minutiflora - Little Gold Poppy

PINACEAE - PINE
Pinus edulis - Two-Needle Piñon
Pinus monophylla - Piñon Pine

POACEAE - GRASS
Bouteloua gracilis - Blue Grama
Hilaria jamesii - Galleta Grass
Hilaria rigida - Big Galleta
Oryzopsis hymenoides - Indian Rice Grass
Panicum urvilleanum - Dune Panic Grass
Stipa speciosa - Desert Stipa

POLEMONIACEAE - PHLOX
Linanthus aureus - Golden Gilia
Loeseliastrum setosissima punctata - Spotted Langloisia
Phlox viridis - Longleaf Phlox

POLYGONACEAE - BUCKWHEAT
Chorizanthe rigida - Rigid Spiny-Herb
Eriogonum deflexum brachypodum - Tecopa Skeleton Weed
Eriogonum fasciculatum - California Buckwheat
Eriogonum inflatum - Desert Trumpet
Eriogonum plumatella - Flat-Top
Eriogonum wrightii - Wright Buckwheat
Oxytheca perfoliata - Punctured Bract
Rumex hymenosepalus - Wild Rhubarb

RANUNCULACEAE - CROWFOOT
Delphinium parishii - Desert Larkspur

ROSACEAE - ROSE
Coleogyne ramosissima - Blackbrush
Cowania mexicana - Cliff Rose
Fallugia paradoxa - Apache Plume
Prunus fasciculata - Desert Almond
Purshia glandulosa - Bitterbush

RUTACEAE - RUE
Thamnosma montana - Turpentine Broom

SALICACEAE - WILLOW
Populus fremontii - Fremont Cottonwood

SCROPHULARIACEAE - FIGWORT
Castilleja chromosa - Indian Paintbrush
Mimulus bigelovii - Bigelow Mimulus
Penstemon eatonii - Eaton Firecracker
Penstemon palmeri - Palmer Penstemon

SOLANACEAE - NIGHTSHADE
Datura wrightii - Sacred Datura
Lycium andersonii - Anderson Lycium
Lycium cooperi - Cooper Lycium
Nicotiana attenuata - Coyote Tobacco
Nicotiana trigonophylla - Desert Tobacco
Physalis crassifolia - Thick-Leaved Ground Cherry
Physalis hederaefolia - Ivy-Leaved Ground Cherry

TAMARIACACEAE - TAMARISK
Tamarix aphylla - Athel
Tamarix ramosissima - Salt Cedar

VERBENACEAE - VERVAIN
Verbena Gooddingii - Goodding Verbena

ZYGOPHYLLACEAE - CALTROP
Larrea tridentata - Creosote Bush

INDEX BY FLOWER COLOR

PLANTS WITH WHITISH OR GREEN FLOWERS

Anderson Lycium
Apache Plume
Arizona Popcorn Flower
Athel
Bitterbush
Blue Yucca
Brown-Eyed Primrose
California Buckwheat
Cooper Lycium
Coyote Tobacco
Desert Almond
Desert Alyssium
Desert Chicory
Desert Holly
Desert Lily
Desert Milkweed
Desert Primrose
Desert Tobacco
Dodder
Dune Primrose
Flat-Top
Forget-Me-Not
Fremont Pincushion
Joshua Tree
Lilac Sunbonnet
Mojave Yucca

New Mexico Thistle
Palmer Penstemon
Pencil Cholla
Pepper Grass
Piñon Brickellia
Prickly Poppy
Punctured Bract
Rattlesnake Weed
Sacred Datura
Sandpaper Plant
Silver Cholla
Spectacle Pod
Spiny Menodora
Squaw Waterweed
Tecopa Skeleton Weed
Tidy Fleabane
Tumbleweed
Water-Wally
White Aster
White Evening Primrose
White Tidy Tips
Winter Fat
Wishbone Bush
Woolly Brickellia
Wright Buckwheat

PLANTS WITH YELLOW FLOWERS

Allscale
Barberry
Barrel Cactus
Big Sagebrush
Bird of Paradise
Black-Banded Rabbitbrush
Blackbrush
Bladderpod
Buckhorn Cholla
Burrobush

California Brickellia
California Croton
Catclaw
Cheesebush
Chinch-Weed
Cliff Rose
Cooper Goldenbush
Cottontop Cactus
Coyote Melon
Creosote Bush

PLANTS WITH YELLOW FLOWERS (Continued)

- Desert Agave
- Desert Dandelion
- Desert Gold Poppy
- Desert Marigold
- Desert Olive
- Desert Senna
- Desert Trumpet
- Four-Wing Saltbush
- Fremont Cottonwood
- Golden Gilia
- Golden Rabbitbrush
- Goldeneye
- Goldenhead
- Grizzly Bear Cactus
- Groundsel
- Ivy-Leaved Ground Cherry
- Linear-Leaved Goldenbush
- Little Gold Poppy
- Live-Forever
- Mariposa Lily
- Matchweed
- Mesquite
- Mojave Horsebrush
- Mojave Prickly Pear
- Mormon Tea
- Old Man Cactus
- Palmer Bead Pod
- Pancake Cactus
- Paperflower
- Piñon Wormwood
- Prince's Plume
- Pygmy Agave
- Rayless Encelia
- Rigid Spiny-Herb
- Rock Goldenbush
- Rock Pea
- Sandwash Groundsel
- Scale Bud
- Silver Puffs
- Tansy Mustard
- Teddy Bear Cholla
- Thick-Leaved Ground Cherry
- Turpentine Brush
- Virgin River Encelia
- Woolly Daisy
- Woolly Marigold
- Woolly-Fruited Bursage
- Yellow Cups
- Yellow Tack Stem

PLANTS WITH BLUE FLOWERS

- Blue Sage
- Canterbury Bell
- Chia
- Desert Hyacinth
- Desert Larkspur
- Indigo Bush
- Mojave Sage
- Smoke Tree

PLANTS WITH PURPLISH FLOWERS

- Beavertail Cactus
- Bigelow Mimulus
- Broomrape
- Chinaberry Tree
- Hedgehog Cactus
- Giant Four O'Clock
- Locoweed
- Mojave Aster
- Notch-Leafed Phacelia
- Paperbag Bush
- Piñon Aster
- Purple Mat
- Range Ratany
- Turpentine Broom
- White Ratany
- Windmills

PLANTS WITH REDDISH FLOWERS

- Eaton Firecracker
- Indian Paintbrush
- Mojave Mound Cactus

PLANTS WITH PINK FLOWERS

Desert Five Spot
Desert Willow
Filaree
Fishhook Cactus
Longleaf Phlox

Pincushion Cactus
Salt Cedar
Sand Verbena
Spanish Needle
Wild Rhubarb

PLANTS WITH ORANGE FLOWERS

Cooper Dyssodia
Desert Mallow

Fiddleneck
Mariposa Lily

GENERAL INDEX

Abronia villosa 185
Acacia greggii 23
Acamptopappus sphaerocephalus 89
Agave deserti 21
Agave utahensis nevadensis 21
Ajo Lily 124
Allionia incarnata 174
Allscale 87
Ambrosia dumosa 77
Ambrosia eriocentra 103
Amsinckia tessellata 180
Anderson Lycium 54
Anisocoma acaulis 147
Antelope Brush 57
Apache Plume 55-56
Apple Brush 110
Apricot Mallow 178
Argemone corymbosa 134
Arizona Popcorn Flower 180
Arrowweed 104
Artemisia ludoviciana 97
Artemisia tridentata 72
Asclepias erosa 125
Aster
　White 142
　Piñon 172
　Mojave 170
Astragalus layneae 168
Astragalus lentiginosus borreganus 168
Athel 22
Atriplex
　canescens 87
　hymenelytra 63
　polycarpa 87
Baccharis glutinosa 67
Baccharis sergiloides 67
Baileya multiradiata 149
Baileya pleniradiata 149
Balloon Flower 133
Banana Yucca 17
Barberry 71
Barrel Cactus 35-36
Bead Pod 158
Beard-Tongue 133
Beavertail Cactus 37
Berberis haematocarpa 71
Big Galleta 52
Big Sagebrush 72-73
Bigelow Mimulus 117, 161
Bird of Paradise 188
Birdcage Primrose 127
Bisnaga 35
Bitterbush 57, 81

Black-Banded Rabbitbrush 74, 118
Blackbrush 75
Bladder Sage 108
Bladderpod 76
Blood Weed 180
Blue
　Dicks 164
　Grama Grass 50
　Sage 105
　Yucca 17-18
Borrego Locoweed 168-169
Bottle Brush 108
Bouteloua gracilis 50
Boxthorn 60
Brickell Bush 78
Brickellia
　californica 78
　incana 69
　oblongifolia 69
Brigham Tea 95
Broomrape 162
Brown-Eyed Primrose 121
Buck Brush 57
Buckhorn Cholla 38-39
Burbush 103
Burrobush 77, 80
Burroweed 77
Burroweed Strangler 162
Bush Encelia 99
Caesalpinia gilliesii 188
California
　Brickellia 78
　Buckwheat 58-59
　Christmas Berry 101
　Croton 79
　Juniper 34
Calochortus kennedyi 184
Calochortus kennedyi munzii 184
Calycoseris parryi 147
Camissonia brevipes 160
Camissonia clavaeformis 121
Canaigre 187
Cane Cholla 38
Canterbury Bell 119, 171
Canyon Live Oak 21
Cassia armata 86
Castilleja chromosa 182
Catclaw 23-24
Cattle Spinach 87
Cedar 33
Celtis sp. 84
Century Plant 21
Ceratoides lanatum 68
Chaenactis fremontii 130

Chamaesyce albomarginata 136
Chamise 87
Cheesebush 80
Chia 163
Chicory 123
Chilopsis linearis 25
Chinaberry Tree 188
Chinch-Weed 145
Chorizanthe rigida 156
Chrysothamnus paniculatus 74
Chrysothamnus nauseosus 90
Cirsium neomexicanum 132
Cleome isomeris 76
Cliff Rose 57, 81
Clocks 181
Coleogyne ramosissima 75
Cooper
   Dyssodia 176
   Goldenbush 91-92
   Lycium 60
Coryphantha vivipara deserti 47
Cottontop Cactus 40
Cottonwood Tree 26
Cowania mexicana 81
Cowboy's Fried Egg 134
Coyote
   Gourd 146
   Melon 146
   Tobacco 122
Creosote Bush 82-83
Croton californicus 79
Cryptantha spp. 180
Cucurbita palmata 146
Cuscuta spp. 129
Cytisus scoparius 86
Daisy, Wallace 159
Dalea fremontii 106
Dalea spinosa 30
Darning Needle Cactus 46
Datura meteloides 137
Datura wrightii 137
Deer Vetch 157
Delphinium parishii 165
Descurainia pinnata 158
Desert
   Agave 21
   Almond 61
   Alyssium 62
   Aster 170
   Baccharis 67
   Buckwheat 58
   Cassia 86
   Catalpa 25
   Chicory 123
   Dandelion 147
   Five Spot 114, 177
   Gold 151
   Gold Poppy 154
   Holly 63
   Hollyhock 178
   Hyacinth 164
   Larkspur 165
   Lily 124
   Mallow 178
   Marigold 149
   Mariposa 184
   Milkweed 125-126
   Needle-Grass 51
   Olive 84-85
   Paintbrush 182
   Primrose 127-128
   Senna 86
   Stipa 51
   Sunflower 88
   Thistle 132
   Thorn 54
   Tobacco 122
   Trumpet 150
   Willow 25
Diamond Cholla 46
Dichelostemma pulchella 164
Dithyrea californica 138
Dock 187
Dodder 129
Dudleya saxosa 155
Dune Panic Grass 9
Dune Primrose 127
Dyssodia cooperi 176
Eaton Firecracker 179
Echinocactus acanthodes 35
Echinocereus engelmannii 41
Echinocereus triglochidiatus 42
Encelia frutescens 99
Encelia virginensis 102
Ephedra nevadensis 95
Ephedra viridis 95
Eremalche rotundifolia 177
Ericameria
   cooperi 91
   cuneata 91
   laricifolia 91
   linearifolia 91
Erigeron pumilus concinnoides 140
Eriogonum
   deflexum brachypodum 139
   fasciculatum 58
   inflatum 150
   plumatella 64
   wrightii 58
Eriophyllum wallacei 159
Erodium cicutarium 181
Eschscholtzia glyptosperma 154

Eschscholtzia minutiflora 154
Euphorbia albomarginata 136
Evening Primrose 127
Fallugia paradoxa 55
Felt Thorn 94
Ferocactus acanthodes 35
Fiddleneck 180
Filaree 181
Firecracker Penstemon 179
Fishhook Cactus 47
Five Spot 114, 177
Flat-Top 64
Fleabane 140
Forestiera neomexicana 84
Forget-Me-Not 180
Four O'Clock 119, 166
Four-Wing Saltbush 87
Fremont Cottonwood 26
Fremont Pincushion 130
Galleta Grass 52
Giant Four O'Clock 119, 166
Globe Mallow 178
Golden
   Cholla 48
   Gilia 151
   Rabbitbrush 90
Goldenbush 91-92
Goldeneye 88
Goldenhead 89
Goodding Verbena 167
Granddaddy Cactus 44
Gray Ball Sage 105
Grayia spinosa 110
Greasewood 82
Great Basin Sagebrush 72
Grizzly Bear Cactus 44
Ground Cherry 152
Groundsel 153
Gutierrezia microcephala 93
Hackberry 84
Haplopappus
   cooperi 91
   cuneatus 91
   laricifolius 91
   linearifolius 91
Hedgehog Cactus 41, 120
Hen and Chickens 155
Heron Bill 181
Hesperocallis undulata 124
Hilaria jamesii 52
Hilaria rigida 52
Holly-Grape 71
Hollyhock 178
Honey Mesquite 23-24
Hyacinth 164
Hymenoclea salsola 80

Indian Paintbrush 117, 182
Indian Ricegrass 53
Indigo Bush 106, 115
Isomeris arborea 76
Ivy-Leaved Ground Cherry 152
Jimson Weed 137
Joint Fir 95
Joshua Tree 15-16, 66
Jumping Cholla 49
June Berry 54
Juniper 33
Juniperus californica 34
Juniperus osteosperma 33
Krameria grayi 109
Krameria parvifolia imparata 109
Langloisia punctata 131
Larkspur 165
Larrea divaracata 82
Larrea tridentata 82
Layia glandulosa 143
Layne Locoweed 168-169
Leoseliastrum setosissima punctata 131
Lepidium fremontii 62
Lepidium lasiocarpum 158
Lesquerella palmeri 158
Leucelene ericoides 142
Lilac Sunbonnet 116, 131
Linanthus aureus 151
Linear-Leaved Goldenbush 91-92
Little Gold Poppy 154
Little-Leaved Ratany 109
Live-Forever 155
Locoweed 168
Longleaf Phlox 183
Lotus rigidus 157
Lycium andersonii 54
Lycium cooperi 60
Machaeranthera canescens 172
Machaeranthera tortifolia 170
Mahonia haematocarpa 71
Malacothrix glabrata 147
Mallow 178
Mammillaria tetrancistra 47
Mariposa Lily 120, 184
Matchweed 93
Melia azedarach 188
Menodora spinescens 66
Mesquite 23-24
Microseris lindleyi 174
Milkvetch 168
Milkweed 125
Mimulus bigelovii 161
Mirabilis bigelovii 144
Mirabilis multiflora 166
Mistletoe 23, 34

Mock June Berry 60
Mojave
  Aster 170
  Horsebrush 94
  Mound Cactus 42
  Prickly Pear 43
  Sage 107
  Yucca 19-20
Monkey Flower 161
Mormon Tea 95
Mound Cactus 42
Nama demissum 173
New Mexico Thistle 132
Nicotiana attenuata 122
Nicotiana trigonophylla 122
Nipple Cactus 47
Notch-Leafed Phacelia 171
Nut Pine 27
Oenothera
  avita 127
  caespitosa 127
  deltoides 127
Old Man Cactus 44
Opuntia
  acanthocarpa 38
  basilaris 37
  bigelovii 49
  chlorotica 45
  echinocarpa 48
  erinacea erinacea 44
  erinacea ursina 44
  phaeacantha 43
  ramosissima 46
Orobanche cooperi 162
Oryzopsis hymenoides 53
Oxytheca perfoliata 135
Paintbrush 182
Palafoxia arida 186
Palmer Bead Pod 158
Palmer Penstemon 133
Pancake Cactus 45
Panicum urvilleanum 9
Paperbag Bush 108, 113
Paperflower 96
Paradise Poinciana 188
Parosela spinosa 30
Peach Thorn 60
Pectis papposa 145
Pencil Cholla 46
Penstemon eatonii 179
Penstemon palmeri 133
Pepper-Grass 158
Petalonyx thurberi 65
Phacelia
  campanularia 171
  crenulata ambigua 171

  fremontii 175
Phlox viridis 183
Phoradendron californicum 23
Phoradendron juniperinum 34
Physalis crassifolia 152
Physalis hederaefolia 152
Pincushion 130
Pincushion Cactus 47
Piñon
  Aster 172
  Brickellia 69-70
  Pine 27-28
  Wormwood 97
Pinus monophylla 27
Plagiobothrys arizonicus 180
Pluchea sericea 104
Plume Buckwheat 64
Poppy 134, 154
Populus fremontii 26
Prickly Pear 43, 45
Prickly Poppy 134
Prince's Plume 98
Prosopis glandulosa 23
Prunus fasciculata 61
Psilostrophe cooperi 96
Psorothamnus fremontii 106
Psorothamnus spinosus 30
Punctured Bract 135
Purple Mat 173
Purple Sage 105
Purshia glandulosa 57
Pygmy Agave 21
Quercus chrysolepis 32
Quercus turbinella 32
Quinine Bush 81
Rabbitbrush 74, 90, 118
Rafinesquia neomexicana 123
Range Ratany 109
Rattlepod 168
Rattlesnake Weed 136
Rattleweed 168
Rayless Encelia 99
Rhus trilobata 101
Rigid Spiny-Herb 156
Rock Goldenbush 91-92
Rock Pea 157
Rosin Weed 93
Rubber Rabbitbrush 90
Rumex hymenosepalus 187
Russian Thistle 141
Sacred Datura 137
Sagebrush 72
Saguaro 35
Salazaria mexicana 108
Salsola australis 141
Salsola iberica 141

Salt Cedar 22, 29
Salvia
  columbariae 163
  dorrii 105
  mojavensis 107
Sand Bunchgrass 53
Sand Verbena 185
Sandpaper Plant 65
Sandwash Groundsel 100
Saratoga 67
Saucer Plant 135
Scale Bud 147-148
Scented Penstemon 133
Scotch Broom 86
Scrub Oak 32
Seep Willow 67
Senecio douglasii 100
Senecio multilobatus 153
Senna 86
Shrub Live Oak 32
Silver Cholla 48
Silver Puffs 147-148
Single-Leaved Piñon 27
Skeleton Weed 139
Skunk Weed 111
Smoke Tree 30
Snakeweed 93
Snapdragon 133
Sore Eye Poppy 178
Spanish Dagger 17, 19
Spanish Needle 186
Spectacle Pod 138
Sphaeralcea ambigua 178
Spiny
  Chorizanthe 156
  Hopsage 110, 118
  Menodora 66
Spotted Gilia 131
Squaw Waterweed 67
Squawberry 101
Squawbush 101
Stanleya pinnata 98
Sticky Rabbitbrush 74
Stipa speciosa 51
Storksbill 181
Sundrops 160
Sweet Sage 57
Tamarisk 29
Tamarix aphylla 22
Tamarix ramosissima 29
Tanglebrush 84
Tansy Mustard 158
Tecopa Skeleton Weed 139
Teddy Bear Cholla 49
Tetradymia stenolepis 94
Thamnosma montana 111

Thick-Leaved Ground Cherry 152
Thistle 132
Thorn Apple 137
Threadleaf Groundsel 100
Tidy Fleabane 140
Tidy Tips 143
Tufted Evening Primrose 127
Tulip 164
Tumbleweed 141
Turbinella Oak 32
Turpentine Broom 111-112
Turpentine Brush 91, 114
Twin Fruit 66
Two-Needle Piñon 28
Utah Juniper 33-34
Verbena 167, 185
Verbena gooddingii 167
Viguiera deltoidea parishii 88
Virgin River Encelia 102
Wallace Daisy 159
Water-Wally 67
Wedgeleaf Goldenbush 91
White
  Aster 142
  Berry Bush 80
  Bursage 77
  Evening Primrose 127-128
  Margined Spurge 136
  Ratany 109
  Sage
  Tidy Tips 143
Wild
  Aster 88
  Heliotrope 171
  Rhubarb 187
Windmills 174
Wingscale 87
Winter Fat 68
Wishbone Bush 144
Wolfberry 54
Woolly
  Brickellia 69
  Daisy 159
  Marigold 149
  Fruited Bursage 103
  Wright Buckwheat 58
Xylorhiza tortifolia 170
Yellow
  Cups 160
  Tack Stem 115, 147-148
  Throats 175
Yucca
  baccata 17
  brevifolia jaegeriana 15
  schidigera 19

Notes